Sacred Vine of Spirits:
AYAHUASCA

ALSO BY RALPH METZNER

Sacred
Vine of Spirits:
AYAHUASCA

EDITED BY

RALPH METZNER, PH.D.

Park Street Press
Rochester, Vermont

Park Street Press
One Park Street
Rochester, Vermont 05767
www.InnerTraditions.com

Park Street Press is a division of Inner Traditions International

LIBRARY OF CONGRESS CATALOGING-IN-PUBLICATION DATA

Sacred vine of spirits : ayahuasca / edited by Ralph Metzner.
    p. ; cm.
  Originally published: New York : Thunder's Mouth Press, ©1999, under the title Ayahuasca.
  Includes bibliographical references.
  Summary: "A compilation of writings on the chemical, biological, psychological, and experiential dimensions of Ayahuasca"—Provided by publisher.
  ISBN 1-59477-053-0 (pbk.)
  1. Ayahuasca—Psychotropic effects. 2. Shamanism—Amazon River Region—Psychology.
  [DNLM: 1. Hallucinogens—therapeutic use—South America—Personal Narratives. 2. Banisteriopsis—South America—Personal Narratives. 3. Ethnopharmacology—South America—Personal Narratives. 4. Indians, South American—South America—Personal Narratives. 5. N,N-Dimethyltryptamine—therapeutic use—South America—Personal Narratives. 6. Phytotherapy—methods—South America—Personal Narratives. 7. Shamanism—South America—Personal Narratives. QV 77.7 S123 1999a] I. Metzner, Ralph. II. Ayahuasca.
  BF209.A93A93 2006
  615.8'80985—dc22
                                                    2005027165

Printed and bound in the United States by Lake Book Manufacturing, Inc.

10   9   8   7   6   5   4   3   2

Text design and layout by Virginia Scott Bowman
This book was typeset in Sabon, Gill Sans, and Agenda with Delphin, Avenir, Schneidler Initials, and Mason Alternate as the display typefaces

To send correspondence to the author of this book, mail a first class letter to the author c/o Inner Traditions • Bear & Company, One Park Street, Rochester, VT 05767, and we will forward the communication.

CONTENTS

This volume is dedicated to

the anonymous men and women,

shamans and curanderas

of the Amazon rain forest,

who preserved and passed on

the knowledge and practices

of ayahuasca, vine of the spirits,

for our benefit and all

sentient beings.

Introduction

Amazonian Vine of Visions

RALPH METZNER, PH.D.

Ayahuasca is an hallucinogenic Amazonian plant concoction that has been used by native Indian and mestizo shamans in Peru, Colombia, and Ecuador for healing and divination for hundreds, perhaps thousands, of years. It is known by various names in the different tribes, including *caapi, natéma, mihi,* and *yagé.* The name *ayahuasca* is from Quechua, a South American Indian language: *huasca* means "vine" or "liana" and *aya* means "souls" or "dead people" or "spirits." Thus "vine of the dead," "vine of the souls," or "vine of the spirits" would all be appropriate English translations. It is however slightly misleading as a name, since the vine *Banisteriopsis caapi* is only one of two essential ingredients in the hallucinogenic brew, the other one being the leafy plant *Psychotria viridis,* which contains the powerful psychoactive dimethyltryptamine (DMT). It is the DMT, derivatives of which are also present in various other natural hallucinogens, including the magic mushroom of Mexico, that provides visionary experiences and thus access to the realm of spirits and the souls of deceased ancestors. DMT is not orally active but is metabolized by the stomach enzyme monoamine oxidase (MAO). Certain chemicals in the vine inhibit the action of MAO and are therefore referred to as MAO-inhibitors: their presence in the brew makes the psychoactive principle available and allows it to circulate through the bloodstream into the brain, where it triggers the visionary access to otherworldly realms and beings. The details of this remarkably sophisticated indigenous psychoactive drug delivery system, and the history of its discovery by science, will be described and explored in this volume.

As a plant drug or medicine, ayahuasca is one of a group of similar substances that defy classification: they include psilocybin derived from the Aztec sacred mushroom *teonanácatl,* mescaline derived from the Mexican and North American peyote cactus, DMT and various chemical relatives derived from South American snuff powders known as *epena* or *cohoba,* the infamous LSD derived from the ergot fungus that grows on grains, ibogaine derived from the root of the African *Tabernanthe iboga* tree, and many others. As plant extracts or synthesized drugs, these substances have been the subject of a large variety of scientific research approaches over the past fifty years, particularly as to their potential applications in psychotherapy, in the expansion of consciousness for the enhancement of creativity, and as amplifiers of spiritual exploration. They have been called *psychotomimetic* ("madness mimicking"), *psycholytic* ("psyche loosening"), *psychedelic* ("mind manifesting"), *hallucinogenic* ("vision inducing"), and *entheogenic* ("connecting to the sacred within"). The different terms reflect the widely differing attitudes and intentions, the varying *set* and *setting* with which these substances have been approached. We will be describing the Western scientific psychological and psychiatric approaches to ayahuasca in this book also.

The concepts of *shaman* and *shamanism* are not peculiar to South America; the terms themselves are derived from a Siberian language. In recent years they have come to be used for any practice of healing and divination that involves the purposive induction of an altered state of consciousness, called the "shamanic journey," in which the shaman enters into "nonordinary reality" and seeks knowledge and healing power from spirit beings in those worlds. The two most widespread shamanic techniques for entering into this altered state are rhythmic drumming, practiced more in the Northern Hemisphere (Asia, America, and Europe), and hallucinogenic plants or fungi, practiced more in the tropics and particularly in Central and South America. Ayahuasca is widely recognized by anthropologists as being probably the most powerful and most widespread shamanic hallucinogen. In the tribal societies where these plants and plant preparations are used, they are regarded as embodiments of conscious intelligent beings that only become visible in special states of consciousness, and who can function as spiritual teachers and sources of healing power and knowledge. The plants are referred to as "medicines," a term that means more than a drug: something like a healing power or energy that can be associated with a plant, a person, an animal, even a place. They are also referred to as "plant teachers" and there are still extant traditions of many-years-long initiations and trainings in the use of these medicines. The use of ayahuasca in the context of Amazonian shamanism is another topic of this book.

Many Western-trained physicians and psychologists have acknowledged that these substances can afford access to spiritual or transpersonal dimensions of consciousness, even mystical experiences indistiguishable from classic religious mysticism, whether Eastern or Western. The new term *entheogen* attempts to recognize this element of access to sacred dimensions and states. In the North American peyote church, the African Bwiti cult using iboga, and in several Brazilian churches using ayahuasca, we have seen the development of authentic folk religious movements that incorporate these entheogenic or hallucinogenic plant extracts as sacraments—developing both syncretic and highly original forms of religious ceremony. The Brazilian ayahuasca-using churches by now have thousands of followers, both in South America and in North America and Europe, and they are growing in numbers and influence. So here we have a substance that has profoundly affected the transformation of individuals now beginning to bring about something like a cultural transformation movement. These facets of the ayahuasca story will also be explored in this book.

As hundreds, perhaps thousands, of Westerners and Northerners have participated in shamanic practices involving ayahuasca (as well as other medicines and nondrug practices) and joined the ceremonies of the various ayahuasca churches, it has become clear that there is a profound discontinuity in fundamental worldview and values between

the Western industrialized world and the beliefs and values of traditional shamanistic societies and practitioners. A powerful resurgence
of respectful and reverential attitudes toward the living Earth and all
its creatures seems to be a natural consequence of explorations with
visionary plant teachers. As such, this revival of entheogenic shamanism can be seen as part of a worldwide response to the degradation of
ecosystems and the biosphere—a response that includes such movements as deep ecology, ecofeminism, bioregionalism, ecopsychology,
herbal and natural medicine, organic farming, and others. In each of
these movements there is a new awareness, or rather a revival of ancient
awareness, of the organic and spiritual interconnectedness of all life on
this planet.

As a psychologist, I have been involved in the field of consciousness studies, including altered states induced by drugs, plants, and
other means, for over forty years. In the 1960s I worked at Harvard
University with Timothy Leary and Richard Alpert, doing research on
the possible therapeutic applications of psychedelic drugs, such as LSD
and psilocybin. During the 1970s the focus of my work shifted to the
exploration of nondrug methods for the transformation of consciousness, such as are found in Eastern and Western traditions of yoga, meditation, and alchemy, and new psychotherapeutic methods using deep
altered states. During the 1980s I came into contact with the work of
Michael Harner and others, who have studied shamanic teachings and
practices around the globe involving nonordinary states of consciousness induced by drumming, hallucinogenic plants, fasting, wilderness
vision questing, sweat lodges, and others. Realizing that there were
traditions reaching back to prehistoric times of the respectful use of
hallucinogens for shamanic purposes, I became much more interested
in plants and mushrooms that have a history of such use, rather than
the newly discovered powerful drugs, the use of which often involves
unknown risks. I have come to see the revival of interest in shamanism
and sacred plants as part of the worldwide seeking for a renewal of the
spiritual relationship with the natural world.

Over the past two millennia Western civilization has increasingly
developed patterns of domination based on the assumption of human
superiority. The dominator pattern has involved the gradual desacral-

ization, objectification, and exploitation of all nonhuman nature. Alternative patterns of culture survived, however, among indigenous peoples, who preserved animistic belief systems and shamanic practices from the most ancient times. The current intense revival of interest in shamanism, including the intentional use of entheogenic plant sacraments, is among the hopeful signs that the split between the sacred and the natural can be healed again.

A recognition of the spiritual essences inherent in nature is basic to the worldview of indigenous peoples, as it was for our own ancestors in preindustrial societies. In shamanistic societies, people have always devoted considerable attention to cultivating a direct perceptual and spiritual relationship with animals, plants, and the Earth itself with all its magnificent diversity of life. Our modern materialist worldview, obsessively focused on technological progress and on the control and exploitation of what are arrogantly called "natural resources," has become more or less completely dissociated from such a spiritual awareness of nature. This split between human spirituality and nature has some roots in the ancient past of Western culture, but a major source of it was the rise of mechanistic paradigms in science in the sixteenth and seventeenth centuries.

As a result of the conflict between the Christian church and the new experimental science of Newton, Galileo, Descartes, and others, a dualistic worldview was created. On the one hand was science, which confined itself to material objects and measurable forces. Anything having to do with purpose, value, morality, subjectivity, psyche, or spirit was the domain of religion, and science stayed out of it. Inner experiences, subtle perceptions, and spiritual values were not considered amenable to scientific study and came therefore to be regarded as inferior forms of reality—"merely subjective" as we say. This encouraged a purely mechanistic and myopically detached attitude toward the natural world. Perception of and communication with the spiritual essences and intelligences inherent in nature have regularly been regarded with suspicion, or ridiculed as misguided "enthusiasm" or "mysticism."

This strange course of events has resulted in a tremendously distorted situation in the modern world, since our own experience, as well as common sense, tells us that the subjective realm of spirit and value is equally as important as the realm of material objects. The revival of animistic, neopagan, and shamanic beliefs and practices, including the sacramental use of hallucinogenic or entheogenic plants, represents a reunification of science and spirituality, which have been divorced since the rise of mechanistic science in the seventeenth century. I believe spiritual values can again become the primary motivation for scientists. It should be obvious that this direction for science would be a lot healthier for all of us and the planet, than science directed, as it is now primarily, toward generating weaponry or profit.

In this book, we will provide a look at the phenomenon of ayahuasca both from the perspectives of objective natural and social science (botany, chemistry, pharmacology, medicine, anthropology, and psychology) and from the point of view of subjective experience—a realm usually not considered amenable to scientific investigation. To do so requires a new look at the epistemology of consciousness.

SCIENCE AND EXPERIENCE— TOWARD AN EPISTEMOLOGY FOR THE STUDY OF CONSCIOUSNESS

Western science in general and psychology in particular have never been comfortable with the study of the subjective side of life: qualities of experience, purposes, intuitions, altered states, or spiritual aspirations. Under the sway of the Newtonian-Cartesian mind-matter dichotomy, consciousness and experience were seen as belonging to the realm of religion, and science agreed to stay out of it. Later, as the ideological hold of the Church diminished and the materialist paradigm became paramount, consciousness and all subjective experience became even more firmly banished from scientific discourse.

In the nineteenth century, the German social philosopher Wilhelm Dilthey attempted to establish the "mental sciences" *(Geisteswissenschaften)* on an equivalent footing to the "natural sciences" *(Naturwissenschaften)*. This idea never really took hold in the English-speaking world. Instead, the social sciences (psychology, sociology, anthropology, political science) adopted and imitated the empirical observational and quantitative analytical methods of the natural sciences. In psychology, the only observations that qualified as scientific were observations of behavior. This was taken to the extreme in B. F. Skinner's behaviorism, in which mental states were said to be in an unknowable "black box." Although the influence of strict behaviorism in psychology has waned in the latter half of the twentieth century, the ideological commitment to a materialist worldview has not. In the leading paradigms of cognitive psychology or cognitive science (which includes brain sciences, computer modeling, information systems, and the like) consciousness is still treated as something to be explained (i.e., explained away) in the supposedly more "real" terms of "neural nets," "brain circuits," and the like.

In the latter half of the nineteenth century a European philosophical movement took a completely different and new approach to the study of consciousness. The German mathematician/philosopher Edmund Husserl (1859–1938) originally conceived of his *phenomenology* as an attempt to rescue philosophy and the quest for absolute knowledge from the naturalism and relativism of the newly arising experimental psychology. He criticized the psychophysical method of Wilhelm Wundt and G. T. Fechner as providing only correlations between subjective events and physical events, and ignoring the possibilities of "pre-understanding," what consciousness was essentially. For Husserl, the abstract truths of mathematics are *essences* that are grasped by the mind directly, without relative or empirical observation. He proposed phenomenology as the method for directly arriving at essential and universal knowledge about the nature of consciousness and meaning, in part by clarifying the implicit pre-understandings that underlie other psychological approaches.

A core concept of Husserl's phenomenology of consciousness is *intentionality:* consciousness is always intentional, always "of" or "about" something, always directed, like an arrow or a mathematical vector, toward some object of meaning. The objects that consciousness intends can be external, or they can be internal aspects of our own experience. Because intentional consciousness is always "constituting" the essential features of the various domains of existence, both external and internal, consciousness has a fundamental "ontological priority"— it is the "supporting ground of reality." The focus on intention as the fundamental constituting attribute of consciousness is congruent with the emphasis on "set and setting" as the prime determinants of altered states. The ontological primacy of consciousness in Husserl's phenomenology is consistent with the worldview of the mystics in Eastern and Western traditions as well as the insights coming from profound altered states.

A further innovative contribution to the phenomenology of consciousness was made by the French philosopher Maurice Merleau-Ponty (1908–1961). In his work, the focus of interest shifts from the subjective mind to the subjective body, or bodily experience *(le corps propre)*. For Merleau-Ponty, perception is an inherently creative, participatory activity between the living body and its world. All subjectivity or consciousness presupposes our inherence in a corporeal world, a world that we perceive as having depth, intimacy, and horizon. The ecophilosopher David Abram (1996) has shown how in many ways Merleau-Ponty's later thought, in his work *The Visible and the Invisible,* anticipates the deep ecologists and others who are looking to develop a new conscious awareness of our embeddedness in the world of nature.

The American philosopher William James (1842–1910) approached the psychology of consciousness in his characteristic multifarious manner. He may have been the first person to use the concept of "field" in talking about consciousness: human beings have "fields of consciousness," which are always complex, containing body sensations, sense impressions, memories, thoughts, feelings, desires, and "determinations of the will," in fact "a teeming multiplicity of objects and relations." He made it clear that his famous "stream of thought" image actually meant not just thoughts, but images, sensations, feelings, etc. He wrote that the mind "seems to embrace a confederation of psychic entities," a statement that contemporary explorers of states of consciousness would readily relate to. In addition to multiplicity, James was greatly impressed by the selectivity of consciousness. In his *Principles of Psychology* he wrote, "The mind is at every stage a theatre of simultaneous possibilities. Consciousness consists in the comparison of these with each other, the selection of some and the suppression of the rest by the reinforcing and inhibiting agency of attention" (James [1890] 1952, 187). The self was the unifying principle in the multiple fields of consciousness, and the active, selective agency that expressed itself through its interests and the directing of attention.

While drawing attention to the multiplicity and selectivity of ordinary consciousness and attention, James also explored the paranormal and mystical dimensions of consciousness that usually lie outside the boundaries of personal or scientific interest. He pursued a lifelong interest in the phenomena of subliminal consciousness, or "exceptional mental states," including those found in hypnotism, automatisms (e.g., somnambulism), hysteria, multiple personality, demoniacal possession, witchcraft, degeneration, and genius. James's interest in unusual states of consciousness led him to experiment with nitrous oxide, or laughing gas as it was then known, an experience that reinforced his understanding of transrational states of consciousness. He wrote that the conclusion he drew from these early psychedelic experiences was "that our normal waking consciousness, rational consciousness as we call it, is but one special type of consciousness, while all about it, parted from it by the filmiest of screens, there lie potential forms of consciousness entirely different" (James [1901] 1958, 228).

James wrote this statement in his *The Varieties of Religious Experience*, probably his most influential book. In it he explored with great discernment and eloquence the nature and significance of mystical or "conversion" experiences, by which he meant not only a person's change from one religion to another, but the process of attaining a sense of unity and the sacred dimension of life. In my book *The Unfolding Self*, I adopted James's empirical, comparative approach to the study of transformative experience—delineating the basic archetypal patterns of psychospiritual transformation. The present collection of accounts of experiences with ayahuasca stands in the same tradition of empirical phenomenology.

It is only recently, in rereading William James's writings on his philosophy of *radical empiricism* (James [1912] 1996) that I came to realize that this philosophy actually provides the epistemology of choice for the study of altered states of consciousness. Within the materialistic paradigm still ruling in scientific circles, any insights or learnings gained from dreams, trances, intuitions, mystical ecstasies, or the like would be seen as "purely subjective" and limited to those states, i.e., not having general applicability or "reality." The altered states of consciousness (ASC) paradigm is still considered marginal. The psychologist Charles Tart in an essay on "state-specific sciences" attempted to break the conceptual stranglehold of this paradigm by suggesting that observations made in a given state of consciousness could only be verified or replicated in that same state. This solution seems theoretically valid, but attended with practical difficulties.

William James started with the basic assumption of the empirical (which means "experience-based") approach: all knowledge is derived from experience. *Die Erfahrung ist die Mutter der Wissenschaft*, as the German saying goes: "experience is the mother of science." James writes:

> I give the name of "radical empiricism" to my *Weltanschauung*.
> . . . To be radical an empiricism must neither admit into its construction any element that is not directly experienced, nor exclude from them any element that is directly experienced. For such a philosophy, the relations that connect experiences must themselves be experienced relations, and any kind of relation experienced must be accounted as "real" as anything else in the system (James [1912] 1996, 42).

This view can provide a philosophical foundation for a scientific psychology of consciousness. All knowledge must be based on observation, i.e., experience; so far this view coincides with the empiricism of the natural and social sciences. It's the second statement that is truly "radical" and that explains why James included religious and paranormal experiences in his investigations. The experiences in modified states of consciousness are currently excluded from materialistic, reductionistic science. They would not be excluded in a radical empiricism.

AYAHUASCA SHAMANISM IN THE AMAZON

The origins of the shamanic use of ayahuasca, as well as other hallucinogenic plants, go back hundreds, perhaps thousands, of years. One cannot really be sure, since we are dealing with cultures that did not keep written records. The anthropologist Geraldo Reichel-Dolmatoff, who spent a lifetime among the indigenous people of Columbia, Peru, and Bolivia, recorded several stories in which the discovery of yagé or ayahuasca is bound up with the origin myth of the people. The Tukano people of the Vaupés region of Columbia say that the first people came from the sky in a serpent canoe, and Father Sun had promised them a magical drink that would connect them with the radiant powers of the heavens. While the men were in the "House of the Waters," attempting to make this drink, the first woman went into the forest to give birth. She came back with a boy radiating golden light, whose body she rubbed with leaves. This luminous boy-child was the vine, and each of the men cut off a piece of this living being that became his piece of the vine lineage. In a variation of this myth from the Desana, the serpent canoe came from the Milky Way, bringing a man, a woman, and three plants for the people—cassava, coca, and caapi. They also regarded it as a gift from the Sun, a kind of container for the yellow-gold light of the Sun, that provided for the first people the rules on how to live and how to speak (Reichel-Dolmatoff 1972).

These origin myths tell us that from the beginning this plant medicine was associated with origins of language, culture, and the beginnings of humankind. Humans are said to have come from the cosmos, and the vine of the soul was given to them as a way of staying in touch with cosmic and solar creative energies. Reichel-Dolmatoff writes for the Indians, "the purpose of taking yagé is to return to the uterus, to the *fons et origo* of all things, where the individual 'sees' the tribal divinities, the creation of the universe and humanity, the first human couple, the creation of the animals, and the establishment of the social order" (102). In conjunction with the cosmic visionary aspects, the yagé mythology and experience among the Indians is saturated with sexual and birth imagery, as well as the shamanic theme of dismemberment. In the Tukano story, the yagé woman who gives birth to the vine-child (which is dismembered) first comes into the house (uterus) and asks the men who the father of the child is. "For the Indian the hallucinatory experience is essentially a sexual one. To make it sublime, to pass from the erotic, the sensual, to a mystical union with the mythic era, the intrauterine state, is the ultimate goal . . . coveted by all" (Reichel-Dolmatoff 1972, 104).

Western explorers in the Amazon region have made sporadic observations of the use of intoxicating plant preparations by the native Indians in the centuries since the conquest. The Catholic priests in the seventeenth century were predictably horrified and condemnatory, as they were of the sacred mushroom cults in Mexico. On the other hand, there were explorers like the German naturalist Baron Alexander von Humboldt in the eighteenth century, and the English botanist Richard Spruce in the nineteenth century, who gave more humanistic and dispassionate accounts of their observations. In the twentieth century, it is above all the work of the eminent botanist Richard Evans Schultes, long-time director of the Botanical Museum at Harvard University, that is responsible for determining the complex ethnobotany of ayahuasca and many other South American psychoactive and medicinal plants. R. E. Schultes's remarkable long life of ethnobotanical explorations and discoveries in the Amazon is the subject of a brilliant and profoundly empathic biography by Wade Davis called *One River* (Davis 1996). The complicated and fascinating history of how ayahuasca was finally correctly botanically identified and its pharmacology analyzed is given in the chapter in this book by Dennis McKenna, who himself contributed crucial pieces of information to the solution of this ethnobotanical puzzle.

In the second half of the twentieth century, increasing numbers of students and researchers in anthropology and ethnobotany were inspired to explore the roots of humankind's involvement with psychoactive plants in shamanism. Starting in the 1950s and 1960s, this cross-cultural research occurred simultaneously with the discovery of psychedelic drugs and their introduction into Western psychology and psychiatry, to be discussed below. These works ranged from R. Gordon Wasson's rediscovery of the pre-Columbian magic mushroom cult and Michael Harner's early work on the role of hallucinogens in European witchcraft-shamanism to the work of sober researchers like Weston LaBarre, Richard Evans Schultes, Claudio Naranjo, and Peter Furst, as well as the more fantastic and imaginative writings of Carlos Castaneda and Terence McKenna.

The shamanic lore of ayahuasca entered most strongly into Western culture initially through the *Yagé Letters* of William Burroughs and Allen Ginsberg, published in 1963; and then through the biography of

Manuel Córdova de Rios, by Bruce Lamb, published in 1971 as *Wizard of the Upper Amazon*. Córdova was abducted as a teenager by a tribe of Indians and initiated into healing knowledge through a lengthy series of ayahuasca sessions in which he got to know the flora and fauna of the rainforest through precise and verifiable visions.

In the Amazon region meanwhile, the shamanic use of ayahuasca, as well as other plants such as tobacco, which is considered one of the most powerful psychoactive and healing plants, moved out of the purely Indian context and into the urban centers with their mestizo populations. The mestizo shamanism involving ayahuasca is known as *vegetalismo* and its practicioners as *vegetalistas*. These healers will use ayahuasca in their curing ceremonies but often know and work with other more straightforward medicinal herbs as well. The gastrointestinal purging reaction is considered essential to the curing, and the yagé brew is therefore often referred to as *la purga*. A book by Luis Eduardo Luna (1986), an anthropologist who has made a special study of vegetalismo, describes the essential features of this shamanic practice.

The traditional shamanic ceremonial form involving hallucinogenic plants is a loosely structured experience, in which a small group of people come together with respectful, spiritual attitude to share a profound inner journey of healing and transformation facilitated by these powerful catalysts. The initiatory training of a healer, as well as certain special healing sessions, may only involve one or two persons besides the elder *ayahuasquero*. A "journey" is the preferred metaphor in shamanistic societies for what psychologists call an altered state of consciousness, or anthropologists nonordinary reality. It is a period of time in which the individual may feel psychically that they are traveling, even flying, or they may feel immersed in strange and sometimes terrifying perceptions that are far from their ordinary experience—all the while the physical body is lying or sitting on the ground with the other participants in the ceremony.

There is a paradox in the terminology often used here to describe these substances. The word "hallucinogen" has been generally rejected by Western psychedelic researchers as being an inappropriate appellation, since they do not induce one to see "hallucinations" in the sense of illusory or nonreal perceptions. But the derivation of hallucination is from the Latin *alucinar,* "to wander in the mind," in other words, an altered state journey. So, I actually prefer to use the term hallucinogen, if it is understood in the sense of "inducing journeys in the mind."

One significant element of virtually all shamanic curing ceremonies involving ayahuasca, and other psychoactive plants and mushrooms as well, is the shaman's singing, which is invariably considered essential to the success of the healing or divinatory process. The singing typical in entheogenic rituals usually has a fairly rapid beat, similar to the rhythmic pulse in shamanic drumming journeys (widespread in shamanistic societies of the northern hemisphere in Asia, Europe, and America). Psychically, the rhythmic chanting, like the drum pulse, seems to give support for moving through the flow of visions and minimizes the likelihood of getting stuck in frightening or seductive experiences. The songs the ayahuasqueros sing are called *icaros* and often have a kind of soft, soothing, almost lilting quality. The songs are learned by the healers in their apprenticeship and are said to be the songs of the spirits that have become the allies of that particular healer. So here is a radically alternative healing system from the accepted Western theory of "magic bullet" drug therapy: the doctor takes the medicine and sings songs, which invoke spirits, who do the healing on the patient.

Another distinctive feature of traditional hallucinogenic ceremonies is that they are almost always done in darkness or low light. This seems to facilitate the emergence of visions, which come with eyes closed from the interior worlds of consciousness, as do dreams. It makes sense that if the visual stimuli of the exterior world are very intense, it would be difficult to pay attention to subtler visual phenomena coming up from within. The exception among psychoactive healing rituals is the peyote ceremony, which is usually done around a fire at night; here participants may see visions as they stare into the fire.

The role of the guide, *curandero,* or healer is always described as central and essential. This must be a person with extensive personal experience in the use of these medicines who agrees to provide an initiatory experience to a seeker or training to an apprentice. In virtually all entheogenic rituals, the guide or shaman does much or all of the singing, and this singing profoundly shapes the quality and content of the experience.

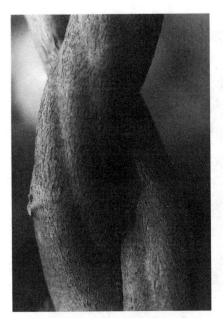 

Banisteriopsis caapi vine
(photo by Jack Coddington)

Psychotria viridis in Hawaiian rain forest
(photo by Ralph Metzner)

The experience can be healing, on physical, psychic, and spiritual levels, although traditional shamanic healers do not make such analytic distinctions. Shamanic healing experiences, with entheogens or other means, have three main variations: the first is the extraction of a toxin that may have been implanted by means of sorcery; the second is the retrieval of a split-off psychic fragment or soul; and the third is the experience of being dismembered or destroyed, and then reconstituted with a healthier, stronger "body."

The experience can provide access to hidden knowledge; this is the aspect of divination, "seeing," prophecy, intuition, or visioning. If the intention or context is healing, then the divination would be equivalent to what Western medicine calls diagnosis—i.e., from where and from whom did the particular toxic implant come, where has the soul-fragment been "lost," what particular herbs should be used for the person's illness, etc. It is said that there is an intelligence associated with the plant medicine, an intelligence that communicates in an interior way to the person who ingests the medicine. Indigenous healers refer to the entheogenic plants as "plant teachers."

Preparation of ayahuasca: pieces of the vine are crushed (to release the chemicals in them) and cooked in a pot of water, in alternating layers with the leaves of *P. viridis.* (photo by Ralph Metzner)

There is a feeling and perception of access to metaphysical realms or worlds, as well as traveling by nonordinary means in this world. In the shamanic traditions, with or without hallucinogens, these realms are called "upper world" and "lower world," as well as "middle world," or more generally "spirit world." In Western esoteric and magical traditions these realms have been referred to as "inner world," "subtle planes," "faerie world," or "otherworld." Some anthropologists, including Michael Harner, refer to them as "nonordinary reality." The access to these otherworlds may come through a kind of journey to that world; perhaps on the back of an animal or carried by a large bird. Alternatively, one may feel that one can see into the spirit world without moving while still aware of the ordinary present world of timespace as well. Scenery and beings of the other world may appear in our world. In any event, the usual boundaries between the worlds seem to become more permeable during such experiences.

The experience may involve the perception of nonmaterial, nor-

The ayahuasca tea being boiled (photo by Ralph Metzner)

mally invisible, spirit beings or entities. Such spirits are recognized as being associated with particular animals (e.g., serpent, jaguar), certain plants, trees or fungi, certain places (e.g., river, rain forest), deceased ancestors, and other nonordinary entities (e.g., extraterrestrials, elves). It can include the experiences of actually becoming or identifying with that spirit (e.g., the experience of becoming a jaguar or a serpent). The healing and divination is experienced as being done by or with the assistance of such spirits, also referred to as "allies," "power animals," "guardians," or "helpers." In some healing rituals, there may also be contact with bad or malevolent spirits that need to be exorcised or neutralized in some way.

The two elements in the shamanic traditions that pose the most direct and radical challenge to the accepted Western worldview are the existence of multiple worlds or realms of consciousness, and the reality of spirit beings. Such conceptions are considered completely beyond the pale of both reason and science within the mindset of the modern world. However, for the many thousands of explorers from North America and

Europe who have used hallucinogenic plants, including ayahuasca, as a shamanic tool for serious consciousness exploration, including those whose experiences are recounted in this book, the recognition of multiple worlds and the reality of spirit beings is becoming quite common.

RESEARCH WITH PSYCHEDELIC DRUGS IN WESTERN PSYCHOTHERAPY

Except for some experimental work with synthetic harmaline (one of the ingredients in the classical ayahuasca combination) as an aid to psychotherapy by the Chilean psychiatrist Claudio Naranjo in the 1960s, ayahuasca itself has not been used as an adjunct to psychotherapy or to psychological research studies until very recently. However, since ayahuasca can be considered to be in the same class of substances as LSD, mescaline, psilocybin, and DMT, it is useful to briefly review the history of research with psychedelic drugs as an aid to psychotherapy and self-exploration. It could be said the various forms of medical and psychiatric treatment, with and without drugs, are the modern, Western equivalent of the traditional shamanic healing system found among indigenous people.

When the fantastically potent mind-altering qualitites of LSD were first discovered at the height of World War II in a Swiss pharmaceutical lab, they were characterized as psychotomimetic and psycholytic. The prospect of unhinging the mind from its normal parameters for a few hours to simulate madness interested a small number of daring psychiatric researchers as a possible training experience. Predictably, this possiblity also intrigued the military and espionage agencies of both superpowers, especially the Americans. Considerable research effort and expense was devoted for about ten years to determining the most effective surreptitious delivery systems to unsuspecting enemy soldiers, agents, or leaders for maximum confusion, disoriention, or embarrassment (Lee and Shlain 1985). Ironically, and fortunately, it was the capacity of LSD to tap into the hidden mystical potentials of the human mind that ruined its applicability as a weapon of war. Rather than making subjects predictably submissive to mind-control programming, LSD had the unnerving propensity to suspend the existing mental programming and thereby release one into awesome worlds of cosmic consciousness. The military was not prepared to have soldiers or espionage agents turn into mystics.

The first research papers that came out of the Sandoz labs, where Albert Hofmann had synthesized LSD and accidentally discovered its astounding properties, described it as bringing about "psychic loosening or opening" *(seelische Auflockerung)*. This was the psycholytic concept that became the dominant model for LSD-assisted psychotherapy in Europe. In psycholytic therapy, neurotic patients suffering from anxiety, depression, obsessive-compulsive or psychosomatic disorders were given LSD in a series of sessions at gradually increasing doses, while undergoing more or less standard analytic interactions using a Freudian perspective (Passie 1997; Grof 1980). The rationale was that through the psycholysis, the loosening of defenses, the patient would become more vividly aware of his or her previously unconscious emotional dynamics and reaction patterns (presumably acquired in early family interactions), and such insight would bring about a resolution of inner conflicts.

The Czech psychiatrist Stanislav Grof, working within this model, made the startling discovery that in such a series (involving increasing doses) there could be an even deeper psychic opening—to birth and prebirth memories. After resolving the conflicts stemming from the Freudian dynamics of early childhood, patients would find themselves reliving the significant sensory-emotional features of their birth experience—patterns to which Grof gave the name *perinatal matrices* (Grof 1985). After passing through birth trauma imagery, frequently associated with sexual and violent content as well, the individual might find him- or herself in transcendent or mystical dimensions of consciousness. The combination of sex and birth imagery with mystical and cosmic experiences found in this research parallels the similar constellation among the Amazonian Indians using yagé.

More or less simultaneously with the psycholytic approach being developed in Europe, the psychedelic model became the preferred approach in Anglo-American psychological and psychiatric circles. The English psychiatrist Humphrey Osmond, who worked in Canada with Abram Hoffer on the treatment of alcoholism with LSD, and who provided Aldous Huxley with his first mescaline experience (immortalized in *The Doors of Perception*), introduced this term in an exchange of letters with Huxley. First used in the treatment of alcoholics, where it was thought to simulate the often life-changing "bottoming out" experience, psychedelic therapy usually involved one or a small number of high-dose sessions, during which the contents of the unconscious mind would be manifested in the form of vivid hallucinatory imagery, leading to insight and transformation (Passie 1997).

The term "psychedelic" was adopted by Timothy Leary, Frank Barron, Richard Alpert, and the Harvard research project, which did one of its first research studies on the production of behavior change in convicts and started publishing the *Psychedelic Review*. Apart from the prison project, Leary's work focused not so much on treatment or therapy, but rather on exploring the possibilities and values of the psychedelic experience for "normals" (mostly graduate students) as well as artists, musicians, poets, and writers, when provided in a relatively unstructured but supportive, homelike setting. The concept of consciousness expansion was introduced for these experiences, which could be usefully contrasted with the contracted, fixated awareness characteristic of narcotic addictions, as well as obsessions and compulsions in general (Metzner 1994). Leary was also responsible for introducing and popularizing what became known as the set and setting hypothesis, according to which the primary determinants of a psychedelic experience are the internal set (intention, expectation, motivation) and the external setting or context, including the presence of a guide or therapist.

The psychological research on psychedelics as well as the psycholytic and psychedelic psychotherapy applications have been well summarized and reviewed by Lester Grinspoon and James Bakalar in their book *Psychedelics Reconsidered* ([1979] 1997). The history of the introduction of LSD and other hallucinogens into American culture with its many extraordinary and unforeseen social and political consequences has been described by Jay Stevens in his book *Storming Heaven* (1987). Leary's own story of these events in which he was centrally involved is told in his own unique, provocative, and tricksterish style in his several autobiographies, most particularly in *High Priest* ([1968] 1995) and *Flashbacks* (1983).

Despite the seeming theoretical and practical differences between the psycholytic and psychedelic approaches, there are a number of significant fundamental conclusions and directions that they share, and that I now summarize. These are all features of psychoactives-assisted psychotherapy that distinguish this modality from other uses of mood-altering drugs such as tranquilizers or antidepressants in which the patient or client takes a pill and goes home.

It is recognized that psychotherapy with hallucinogens invariably involves an experience of a profoundly expanded state of consciousness in which the individual can not only gain therapeutic insight into neurotic or addictive emotional dynamics and behavior patterns but may come to question and transcend fundamental self-concepts and views of the nature of reality. This understanding is completely congruent with the views of the ayahuasca-using shamans, who say that the medicine can give them insight into their own makeup and show them a better way to live.

It is widely accepted in the field that set and setting are the most important determinants of experiences with psychedelics, while the drug plays the role of a catalyst or trigger. This is in contrast to the psychiatric or other psychoactive drugs, including stimulants, depressants, and narcotics, where the pharmacological action seems paramount, and set and setting play a minor role. The set-and-setting model can also be extended to the understanding of other modalities of altered states of consciousness, involving nondrug triggers such as hypnosis, meditation, rhythmic drumming, sensory isolation, fasting, and others (Metzner 1989). Acceptance of the paramount importance of set and setting is obvious and implicit in the shamanic approach to hallucinogens. A ritual structure is essentially the conscious arranging of the set and setting for the purposes agreed upon by the shaman and the patient or initiate.

Two analogies or metaphors for the drug experience have been repeatedly used by writers both in the psycholytic and psychedelic paradigms. One is the amplifier analogy, according to which the drug functions as a nonspecific amplifier of psychic contents. The amplification may occur in part as a result of a lowering of sensory thresholds, an opening of the doors of perception, and may in part be due to as yet not understood central processes involving one or more neurotransmitters. The other analogy is the microscope metaphor: it has repeatedly been said that psychedelics could play the same role in psychology as the microscope does in biology—opening up realms and processes of the human mind to direct, repeatable, verifiable observation that have hitherto been largely hidden or inaccessible. Both amplifier and microscope are technological metaphors for expanded perception and divination—the ability to see and hear more vividly, to see into other, normally invisible worlds or dimensions, and to obtain otherwise hidden knowledge.

Again in contrast to the use of other psychiatric or psychoactive drugs, it is widely recognized that the personal experience of the therapist or guide with psychedelics is an essential element of effective psychedelic psychotherapy. Without such prior personal experience, communication between the therapist and the individual in a psychedelic state is likely to be severely limited. This principle implies also that a significant role for psychedelic experience could be in the training of psychotherapists. This principle is so self-evident in the shamanic context that it is not even mentioned. Shamanic healers undergo months or sometimes years of training involving personal experience under the guidance of an experienced ayahuasquero, before they would start to heal someone else.

Psychedelics provide access to transcendent, religious, or transpersonal dimensions of consciousness. That mystical and spiritual experiences can and do often occur with psychedelics was recognized early on by most researchers in this field, thereby posing both challenge and promise to the psychological disciplines and professions. Albert Hofmann has testified that his ability to recognize the psycholytic properties of the LSD experience was based on its similarity to his childhood mystical experiences in nature (Hofmann 1979). Stanislav Grof found that after resolving biographical childhood issues, and then the perinatal trauma, individuals would often find themselves in realms of consciousness completely transcendent of time, space, and other parameters of our ordinary worldview (Grof 1985). He gave the name *transpersonal* to these realms of consciousness and *holotropic* ("seeking the whole") to the predominant quality of consciousness in these realms, as well as to other means of accessing these realms, such as certain breathing methods (holotropic breathwork). Timothy Leary, stimulated no doubt by his association with Aldous Huxley, Huston Smith, and Alan Watts, devoted considerable time and energy to exploring and describing the spiritual and religious dimensions of psychedelic experience. These interests resulted in adaptations of the Tibetan Buddhist text *Bardo Thödol* and the Chinese Taoist *Tao Te Ching* as guidebooks for psychedelic experience (Leary, Metzner, and Alpert 1964; Leary 1966/1997).

Thus we find that Western psychologists doing research and psychotherapy with psychedelics came to similar conclusions about the nature of these experiences as did the shamanic explorers in the Amazon or ancient Mexico. In both approaches there is a recognition of the possibility of healing and resolution of personal problems and difficulties, especially deep-seated patterns of guilt and fear around sexuality and birth. Furthermore, and more profoundly, these experiences provide access to the most transcendent and profoundly mystical realms of human experience, teaching us lessons about our most ancient past, our origin, our involvement in the spiritual realms, and our possible future.

A review of what is presently known to science of the psychological effects of ayahuasca and its chemical ingredients is provided in this book in the chapter by Charles Grob. The accounts of subjective experiences with ayahuasca by (mostly) North Americans and Europeans, given in this book, also illustrate and confirm many of the themes and images found among Indian and mestizo shamanic users, as well as among many psychotherapeutic users of various psychedelics.

PLANT ALKALOIDS, NEUROTRANSMITTERS, AND THE HUMAN BRAIN

When scientists have attempted to understand and explain the action of hallucinogenic drugs and plants such as ayahuasca in terms of current models in pharmacology and neurochemistry they have encountered an intriguing challenge as well as a profound mystery. I remember when I did my postdoctoral fellowship in psychopharmacology at the Harvard Medical School in the early 1960s that although there were literally thousands of studies of the action of LSD and other psychoactive drugs on various human and animal physiological and biochemical processes there was not even an inkling of an explanation for the profound psychic effects of these substances. The one study that I recall making some intuitive sense was an investigation into where the metabolic residues of LSD were concentrated in a monkey's brain. The highest concentrations were found in the visual pathways and retina, as well as the pituitary and pineal gland. This finding was consistent with the intense visual hallucinatory effect of the drug, as well as the effects on central regulatory neural and hormonal processes. We know that the pineal gland is deeply involved in the sleep and wakefulness cycle through its production of the hormone melatonin, which is quite closely chemically related to the psychoactive tryptamines DMT and 5-methoxy-DMT.

The one advance in brain biochemistry that has added significantly to our understanding since those early days is the discovery of *neurotransmitters,* substances produced in the body that are released at the junctions (called synapses) between neurons and facilitate the transmision of an electrical signal across the synaptic junction or cleft. Dozens of neurotransmitters have now been identified, but the four that seem to play the most central and pervasive roles are dopamine, epinephrine, acetylcholine, and serotonin. Most of the drugs produced by the pharmaceutical industry to affect mood states—anxiety, depression, psychosis, and the like—affect one or more of these four neurotransmitters, either increasing or inhibiting their action. Of these, it is serotonin, the chemical name of which is 5-hydroxytryptamine, that is generally regarded as the key substance involved in the activity of psychoactive drugs. Serotonin is synthesized in the body from its dietary precursor tryptophan, one of the eight essential amino acids.

The chapter in this book by neurochemist J. C. Callaway presents what is presently known about the complex interactions of the ayahuasca brew with the endogenous serotonin in the human brain, knowledge to which he himself has also contributed through original research. Suffice it to say here that some intriguing pieces of this biochemical puzzle have come to light in the last decade or so. Among them are the fact that serotonin deficiency has been implicated in depression, anxiety, irritability, violence, insomnia, and several other psychological and neurological disorders. This is the basis for the treatment of depression by (dietary or supplemental) tryptophan, as well as by the selective serotonin reuptake inhibitors (SSRIs), such as Prozac, which increase the availability of serotonin in brain circuits. Serotonin is normally catabolized in the body by MAO, and therefore the MAO-inhibitors, such as the harmala alkaloids present in ayahuasca, would lead to an increase in endogenous serotonin levels that may account for the fact that ayahuasca takers are often remarkably calm and unafraid in view of the sometimes terrifying visions that they are confronting. Serotonin is also present in the intestines where it increases intestinal motility, and at high levels can induce vomiting and diarrhea. These are the basis for the purging action of ayahuasca, which is also known among the mestizo healers as *la purga*. While this purging action is usually experienced by the ayahuasca user, at moderate levels, as healing, clearing, and liberating (as can be seen in many of the accounts in chapter 4 of this book), when excessive it could lead to acute physiological distress symptoms. These are regarded as indicative of a "serotonin syndrome," a serotonergic reaction to excessive levels of serotonin. Callaway rightly cautions that the combination of SSRIs and the MAO-inhibiting harmala alkaloids present in ayahuasca could trigger this kind of reaction.

What is truly amazing is that the Indian and mestizo shamans of the Amazon rainforest appear to have figured out or discovered these plant and biochemical interactive effects using only careful naturalistic observation and experimentation. There are variations in the length to which the mixture is boiled, as well as in what other herbs and plants are added to the brew, such as datura leaves or medicinal herbs appropriate to the condition of the person being cured. But there is always at minimum the combination of the two plants, one containing the hallucinogenic tryptamine alkaloids and the other containing the serotonin-elevating and MAO-inhibiting harmala alkaloids. The Quechua name for the tryptamine-containing leafy plant is *chacruna,* and for the vine *mariri.* Among the members of the hoasca church UDV, to be described below, the chacruna is said to bring "light," i.e., visions, and the mariri is said to symbolize "strength." It will be noted that in the subjective experience accounts people do feel strong, even with the seemingly terrifying visions and the purging, it is a strength coming from being in touch with the deepest biological roots of one's being. They understand that it is the harmine-containing vine that produces the purging reaction, so they increase the proportion of that when they want to help the patient purge parasites or other toxins from the body. At the same time, they also seem to understand the possible danger of the serotonergic reaction, since they require the initiate taking frequent doses of ayahuasca to go on a restricted diet, containing minimal levels of tryptophan (usually found in carbohydrates).

Some of the indigenous healers and herbalists are veritable walking encyclopedias of medicinal botanical knowledge. They may have direct personal knowledge of hundreds, even thousands of plants, and what illnesses or conditions they can be used to cure; this knowledge was not acquired by literate means, but by direct experience. "Experience is the mother of science." In addition they have developed a sophisticated understanding of the biochemistry of plant-drug interactions and their effects in the human body. Normally, the MAO enzyme functions as a protective biochemical gate or screen, catabolizing alkaloids that could have a potentially toxic effect on the body. There is evidence that the indigenous herbalists use the harmala containing plants (such as the caapi vine) to test the effects of various plant medicines in their own body by taking low doses that will not be toxic to them. This is also the reason why it is valid to consider the plant preparations used by the indigenous healers, when prepared and taken in the way and with the rituals they use, to be safe. With all the extensive experimentation they have done, going back hundreds of years, these people know enough to distinguish poisons from medicines. They are adepts at plant poisons also—as with the poison blow dart containing curare. This safety factor would not necessarily apply however to mixtures of traditional plants with other substances, much less to novel compounds synthesized in the laboratory that do not occur in nature and have not been tested for long-term effects.

In recent years, inspired by the discovery of the ayahuasca story, a number of botanists, chemists, pharmacologists, and ecologists have begun looking for other plant species, especially those containing psychoactive tryptamines and those containing MAO-inhibiting ß-carbolines, that could be combined to create an ayahuasca-like potion. For example, the leaves of *Psychotria viridis* may be combined with the seeds of Syrian rue *(Peganum harmala)*, which contain harmaline, to create what some call an "ayahuasca analogue" (Ott 1994). Some of these scientists are trained professionals, others are self-taught aficionados of the plant realm; most are not affiliated with mainstream universities or research institutes. Their findings are often written up in self-published books, in unorthodox journals such as the *Entheogen Review*, and to an increasing degree, on the Web (www.lycaeum.org; www.erowid.org). By now, dozens of plants containing ß-carbolines and well over a hundred containing tryptamines such as DMT in varying concentrations have been identified. It should be remembered that the vast majority of these novel ayahuasca-like combinations have never been used by any indigenous tribes for shamanic purposes, so the element of uncertainty regarding safety and toxicity is quite high.

The one exception that is worth mentioning is the plant concoction called *jurema*, which is used by some indigenous tribes in northeastern Brazil. The plant used is a species of *Mimosa* in which the root bark contains large amounts of DMT. It is not known what plants the Indians may have used to deactivate the MAO enzyme. However, Western explorers have used the seeds of Syrian rue for this purpose, combining them with the Mimosa, to yield a combination that is powerfully hallucinogenic, but without the purgative effect of classical ayahuasca. As far as the present volume is concerned, we will concentrate entirely on the classic ayahuasca or yagé experience and knowledge. Readers who wish more information about ayahuasca analogues, as well as other psychoactive or entheogenic plants and combinations, are advised to consult the writings of Jonathan Ott (1993, 1994) or the recently published *The Encyclopedia of Psychoactive Plants* (Rätsch 2005).

Scientists associated with the giant pharmaceutical corporations have long realized that the tropical rainforests of Amazonia and elsewhere are veritable treasure houses of medicinal plants, the large majority of them as yet unknown to medicine. As has often been pointed out, the cure for cancer and other diseases may yet lie undiscovered in rain forest plants. Extensive research projects have been mounted with the aim of finding new plant drugs to test, which would then be isolated, synthesized, patented, and marketed. There is, however, another strand of research activity that starts with collecting the indigenous peoples' ethnobotanical knowledge, recognizing the importance of taking into account the whole pattern of use, the rituals, mythology, and lore of these plants, instead of only the molecular compound that can be isolated from them. Here there is variation and considerable controversy over the extent to which the scientists from the industrialized world will acknowledge and compensate the indigenous people for the medicinal and scientific knowledge they have received. Among this group of scientists too are those who, like the scientists and observers represented and cited in this book, have taken the claims of the indigenous healers who work with hallucinogenic plants seriously, and therefore seek to preserve, in holistic fashion, the entire body of knowledge, from the botanical to the psychological, the cosmic and the spiritual.

For all their demonstrated knowledge of herbs and medicine, the ayahuasqueros are unanimous in their assertion that the knowledge is given to them by the spirits of the plants, the forest, or the animals. Likewise, the healing is done, not so much by the plant drug, but by the spirit or essence invoked by the healer, via the use of the plant teacher, and expressed in the songs. With this belief, which is completely at variance with the accepted medical model focused on isolating and purifying the molecular compound, they would agree with Samuel Hahnemann, the great eighteenth-century German physician who founded homeopathy. In this medical system, the plant drug extracts are repeatedly diluted to such a degree that often not a single molecule of the original substance is left. In addition they are shaken or vibrated, a process referred to as *succussion*. Hahnemann said that through the repeated dilutions and succussions the spirit or essence of the plant was entirely released, or liberated, from the plant substance, and was thus able to act on the spiritual or essence of the patient. In this recognition of the spiritual essences inherent in plant medicines and their healing virtue, the homeopaths and the shamanic healers are in accord. It is also, I would add, the underlying assumption and understanding that I and my colleagues and collaborators have come to and that is represented, explicitly or implicitly, in the accounts in this book.

THE COSMIC SERPENT AND THE DNA—
THE WORK OF JEREMY NARBY

A breakthrough contribution to an integrated scientific understanding of ayahuasca and the shamanic way of knowledge was made by Jeremy Narby, a Canadian-born anthropologist and conservationist who lives in Switzerland. In his book *The Cosmic Serpent* (1998) he recounts how his experiences with ayahuasca shamans among the Ashininca in the Peruvian Amazon led him to a reexamination of the foundations of molecular biology, and eventually to an hypothesis that reconciles the teachings of the ayahuasqueros with findings of modern science. Narby was fully aware of the seemingly irreconcilable conflict between the shamanic worldview, which says that one could obtain reliable knowledge, applicable in healing, from the visions induced by ingesting hallucinogenic plants, and the scientific worldview, which says that the ayahuasca visions are delusional hallucinations caused by plant toxins. He started his journey toward reconciling these opposite perspectives by deciding to take the ayahuasqueros at their word—that valid medicinal knowledge could be obtained from these plant teachers. This decision was strengthened by his own experiences with ayahuasca. We could say, in other words, that he practiced the method of radical empiricism—basing his quest for knowledge on his own experience, and not excluding it because it didn't fit with prevailing theories.

Visions of serpents, sometimes gigantic, sometimes luminous, are extremely common in ayahuasca experiences, as Narby had found in his own experience as well (and as the accounts in this book will confirm). The ayahuasqueros told Narby that the serpent spirit is the mother of ayahuasca, the source of knowledge and healing power. Images of serpents, dancing, moving, double or multiple are found throughout the hallucinatory art work of the Indians. When Narby then started reading the voluminous literature on molecular biology for clues as to how the brain might be affected in visionary states of consciousness, he came to the realization that the DNA molecule, which has the form of a twisted double helix, might be the molecular counterpart to the hallucinated serpents of ayahuasca. Not only among Amazonian shamans, but throughout the world, in Asia, the Mediterranean, Australia, serpent images are used to represent the basic life force and regarded as a source of knowledge—the wisdom of the serpent. The serpent image is seen often as a link between heaven and earth, and in this regard the snake is often found in association with other images of ascent. As Narby writes, "in the literature of molecular biology, DNA's shape is compared not only to two entwined serpents, but also, very precisely, to a rope, a vine, a ladder, or a stairway" (p. 93).

DNA is the molecular code for all life on this planet, whether animal, plant, or human. It is present in every cell of every body, of every plant, animal, fungus, or human. Perhaps, Narby proposed, shamans in their visions manage to take their consciousness down to the molecular level, reading information about how to combine MAO-inhibitors with brain hormones, how to recognize correspondences between certain healing plants and certain illnesses, and so forth. The biological description of the DNA molecule shows startling correspondences with the insights of the shamans, which they claimed to get from their serpent visions, and the mythology of cosmic serpents.

> If one stretches out the DNA contained in the nucleus of a human cell, one obtains a two-yard-long thread only ten atoms wide. . . . The nucleus of a cell is equivalent in volume to two-millionths of a pinhead. The two-yard thread of DNA packs into this minute volume by coiling up endlessly on itself, thereby reconciling extreme length and infinitesimal smallness, like mythical serpents. The average human being is made up of one hundred thousand billion cells, according to some estimates. This means that there are approximately 125 billion miles of DNA in a human body. . . . Your personal DNA is long enough to wrap around the Earth five million times (p. 87–88).

Talk about world-encircling serpents! Furthermore, the DNA molecular code has remained unchanged since the beginning of life on this planet—only the arrangements of the "letters" in the code changes in the evolution of different species. "DNA is the master of transformation, just like mythical serpents. The cell-based life DNA informs made the air we breathe, the landscape we see, and the mind-boggling diversity of living beings of which we are a part" (p. 92). Thus Narby proposed that the DNA double, twisting serpent was the source of shamanic knowledge, whether acquired through ayahuasca or through other altered state techniques such as drumming, fasting, isolation, or dreaming. Furthermore, the fact that the DNA molecule emits biophotons could account for the luminosity of shamanic visions when this molecular luminosity is allowed to emerge by reducing external illumination. Narby's hypothesis, while not offering any new facts or proofs, has revolutionary implications in that it reconciles the diametrically opposite worldviews of science and that of shamanism.

SYNCRETIC FOLK RELIGIOUS CEREMONIES INVOLVING AYAHUASCA

The distinction I have drawn between entheogen-based shamanic rituals and folk religious ceremonies involving plant entheogens is in some ways arbitrary. There is a continuum of ritual forms and practices. The emphasis in shamanic practices is healing and divination, and they are usually conducted in small groups of about a dozen participants, or sometimes just with one or two afflicted individuals and shaman apprentices. The folk religious ceremonies, such as those of the Native American Church or the African Bwiti cult, usually involve fairly large groups of twenty to forty participants, or in the case of the Brazilian ayahuasca churches up to several hundred. In such ceremonies the intentional focus is not so much on healing and visioning, but on group worship and celebration with singing and prayer. Instead of a shaman or healer there are priests and officiants. There is very little or no discussion or sharing of visions or insights, as there would be in the context of a shamanic healing or divination.

The groups coalescing around such entheogenic folk ceremonies in an urban or village society have organized themselves into recognized churches, thereby providing their members with a certain degree of social cohesion and protection. An important social function of these religious ceremonies is to strengthen community bonds and give members a sense of participation and belonging. As Charles Grob reports in his account of the research with the long-term *hoasca* users of the *União do Vegetal*, there is marked reduction in the incidence of alcoholism and drug addiction; this is also true for participants in the Native American Church in the United States. Anthropologists have noted that a further societal function of these churches is to provide a protective shield of traditional lore against the encroachments of Christian missionaries and the seductions of Western consumer culture.

In Brazil there are no less than three organized churches in which ayahuasca is the main sacrament, the *Santo Daime*, the *União do Vegetal* (UDV), and the *Barquinia*. Santo Daime, the oldest church, was founded by Raimundo Irineu, a rubber tapper of African descent, in the 1930s, in the state of Acre. Ingesting ayahuasca with mestizo shamans, he had a vision of a female deity, the Queen of the Forest, who instructed him to create a religious movement that incorporated the tea, which they call *daime* ("give me"), as well as the singing of songs channeled in ayahuasca sessions. The UDV, presently the largest of the groups, was founded independently by another rubber tapper, Gabriel de Costa, in Rondonia state in 1954, who also had a vision of a church; their name for the tea is *hoasca*, which means simply "the vine." Barquinia was also started in the 1950s by ceremonial leaders who broke off from Santo Daime to found a church that incorporated many elements of the Brazilian spiritist religion known as Umbanda. Members of the churches come from all walks of life and both urban and rural environments in Brazil. Each has by now several thousand members in Brazil, and two of the churches have significant satellite centers in North America and Europe. The churches are officially recognized and the use of ayahuasca is legal in Brazil within that framework.

Each of the churches has some differences of emphasis and ceremonial form, although each incorporates some Christian elements into their ceremonies as well. The UDV, the largest, is also the most formal: participants sit in rows in straight-back chairs during the ceremony, listening to sermons and songs given by the *mestres* who sit around a table in the center; there is also a question and answer period. Sometimes testimonials of life transformations are offered by longer-term members, reminiscent of AA recovery confessions. The Santo Daime service involves the singing of hymns by the entire congregation, usually led by a small group of women singers. Some of the Santo

Daime ceremonies also involve dancing, in simple rhythmic, swaying steps—the whole somewhat reminiscent of Black American gospel services (which is interesting in light of the fact that it was founded by a man of African descent). The Barquinia church also uses singing, dancing, and mediumistic contact with *caboclos,* the spirits of deceased Indians and *pretos velhos,* the spirits of deceased Black slaves. A few of the accounts in this book are from North American participants in the Santo Daime or the UDV church.

These syncretic religious movements in Brazil have brought the use of entheogenic plant substances out of the context of shamanic healing rituals where only a very limited number of people came into contact with them. They have made profound spiritually transforming experiences with entheogenic plant medicines accessible to a large number and wide spectrum of people in all walks of life, both in Brazil and also in North America and Europe. These movements represent an authentic religious revitalization movement. We may be seeing the beginnings of a broader transcultural transformation movement with significant impact.

AYAHUASCA TOURISM

Not only have Westerners and Northerners participated in the ayahuasca religious ceremonies of the Santo Daime and the UDV, but there has also been in the past ten years or so a dramatic increase in the number of Europeans and North Americans who have journeyed to the Amazon to participate in guided tours of rain forest ecology and indigenous culture—tours that will often include one or more ayahuasca sessions guided by a local mestizo shaman. The increasing interest in ayahuasca tourism was triggered in part by the writings of the visionary philosopher Terence McKenna and his scientist brother Dennis, especially their astonishing work *The Invisible Landscape* (1975), and Terence McKenna's *True Hallucinations* (1993), books that described their exotic hallucinatory, ethnobotanical, and alchemical adventures. Interest in ayahuasca was further increased by the two books by Bruce Lamb on Manuel Córdova de Rios, the already mentioned *Wizard of the Upper Amazon* (1971) and the sequel *Rio Tigre and Beyond* (1985);

and an amazing collection of paintings of ayahuasca visions by the erstwhile ayahuasquero Pablo Amaringo, along with his story of giving up the practice of healing because of his distaste for the sorcery associated with it (Amaringo and Luna 1991). The rock musicians Sting and Paul Simon each shared stories about their experiences with ayahuasca, which lent further celebrity allure (Grunwell 1998). Several mestizo ayahuasqueros have also traveled repeatedly to the United States and Europe, conducting ceremonies in Western houses with middle-class, educated, mostly white participants. Ayahuasca jungle tours are advertised on the Internet and in magazines such as *Shaman's Drum*.

Ayahuasca tourism has come in for its share of criticism from those who decry naive and possibly exploitative intrusions into indigenous cultures, and those who warn tourists of being duped by ignorant fake shamans, or damaged and poisoned by sorcerers. Nevertheless, one cannot simply dismiss a whole category of people, many of whom are sincerely seeking some spiritual wisdom and insight into their lives and some of whom are desperately seeking relief from illnesses that Western medicine has been unable to cure. Some miraculous cancer "cures" have been reported (MAPS 1998). On the side of the indigenous cultures, the ayahuasca tourists, like other tourists, contribute desperately needed funds to local South American economies. Ecological preserves are being set up to protect the rain forest and all its medicinal and psychoactive plants. Ayahuasca tourism and ecotourism converge in their goals and methods. There is always the possibility of abuse, but the positive effects are also real.

Similarly, the ayahuasca churches have been criticized as exploitative misappropriations of shamanic methods and teachings. It is true the churches have each created their own mythology, and rarely acknowledge the contribution of the native Indians to their medicinal and ethnobotanical practices. Indeed, the hierarchy of the UDV, for example, claim that their founder Mestre Gabriel realized that in a previous incarnation he was an Inca king who showed the Indians how to prepare and use the ayahuasca. Once again, the native people's knowledge and methods are disparaged and marginalized. However, it is not true that the churches represent a decadent shamanic practice. They represent a syncretic religious form that makes this hallucinogenic healing brew available to thousands of urban residents both in South America and in North America and Europe. In meeting with people of the different churches, I have not been drawn to join their particular religion or accept their ideology, but nevertheless I've appreciated their values, which are humane and supportive of family and community.

Out of studies by North Americans in Peru and Brazil, with mestizo shamans and anthropologists who have studied with them, has grown a network of Western psychedelic seekers who come to ayahuasca often with considerable experience with psychotherapeutic uses of psychedelic drugs. It is from this loose collection of consciousness explorers that most of the accounts in this book are drawn. I call their approach a hybrid of shamanic and psychotherapeutic methods.

HYBRID SHAMANIC PSYCHOTHERAPEUTIC RITUALS

In the past dozen years or so I have been a participant and observer in over one hundred circle rituals, in both Europe and North America, involving several hundred individuals, many of them repeat participants. Plant entheogens used in these circle rituals have included psilocybe mushrooms, ayahuasca, San Pedro cactus, iboga, and others. My interest has focused on the nature of the psychospiritual transformation undergone by participants in such circle rituals (Metzner 1998).

In these hybrid therapeutic-shamanic circle rituals, certain basic elements from traditional shamanic healing ceremonies are usually kept intact: the structure of a circle, with participants either sitting or lying; an altar of some kind in the center of the circle, or a fire in the center if outside or in a tipi; the presence of an experienced elder or guide, sometimes with one or more assistants; preference for low light, or semidarkness (sometimes eyeshades are used); the use of music such as drumming, rattling, singing, or evocative recorded music; the dedication of ritual space through invocation of spirits of four directions and elements; and the cultivation of a respectful, spiritual attitude.

Experienced entheogenic explorers understand the importance of set and therefore devote considerable attention to clarifying their intentions with respect to healing and divination. They also understand the importance of setting and therefore devote considerable care to arranging a peaceful place and time, filled with natural beauty and free from outside distractions or interruptions.

Most of the participants in circles of this kind that I witnessed were experienced in one or more psychospiritual practices, including shamanic drum journeying, Buddhist *vipassana* meditation, tantric yoga, holotropic breathwork, and most have experienced and/or practiced various forms of psychotherapy and body-oriented therapy. The insights and learnings from these practices are woven by the participants into their work with the entheogenic medicines. Participants tend to confirm that the entheogenic plant medicines combined with meditative or therapeutic insight processes amplify awareness and sensitize perception, particularly somatic, emotional, instinctual, and spiritual awareness, as well as often a sense of the interconnections between these levels of consciousness.

Some variation of the talking staff or singing staff is often used in such ceremonies. With this practice, which seems to have orginated among the Indians of the Pacific Northwest, and is also more generally referred to as "council," only the person who has the circulating staff sings or speaks, and there is no discussion, questioning, or interpretation as there might be in the usual group psychotherapy formats. Some group sessions, however, involve minimal or no interaction between the participants during the time of the expanded state of consciousness.

In preparation for the circle ritual, there is usually a sharing of intentions and purposes among the participants, as well as the practice of meditation, or sometimes solo time in nature, or expressive arts modalities, such as drawing, painting, or journal work. After the circle ritual, sometimes the morning after, there is usually an integration practice of some kind, which may involve participants sharing something of the lessons learned and to be applied in their lives.

The majority of Westerners who have developed an ongoing practice of working with entheogenic plant substances seem to have expanded their belief systems beyond the boundaries of the conventional materialistic paradigm of Western science and psychology. While accepting the validity of many Western psychological insights, including those of Sigmund Freud, Carl Gustav Jung, and Wilhelm Reich, they have come, like indigenous people, and Asian and Western esoteric traditions, to accept the reality of nonmaterial spirit beings and to recognize that we live in multiple worlds of consciousness. I will discuss the worldview and longer-range implications of these research findings in the concluding chapter of this book.

A common theme found repeatedly in the sessions is a heightened awareness and concern for the protection and integrity of the Earth and its threatened habitats and wildlife, as well as of indigenous cultures. In that sense, it can be said that the growing interest in shamanism in general, and the ayahuasca shamanism in particular, represents part of a worldwide movement toward a more direct, experiential, and spiritual connection to the natural world. The fact that Westerners will seek out a foul-tasting jungle medicine, in a faraway environment and culture, a medicine that frequently leads to violent purging and can include terrifying visions, is a remarkable paradox, and yet the overwhelming majority of people who try it find in it the deepest spiritual realizations of their life as well as profoundly healing changes.

References

Abram, D. 1996. *The Spell of the Sensuous*. New York: Pantheon Books.

Amaringo, P., and L. E. Luna. 1991. *Ayahuasca Visions—The Religious Iconography of a Peruvian Shaman*. Berkeley, CA: North Atlantic Books.

Burroughs, W. S., and A. Ginsberg. 1963. *The Yagé Letters*. San Francisco: City Lights.

Davis, Wade. 1996. *One River—Explorations and Discoveries in the Amazon Rainforest*. New York: Simon & Schuster.

Dobkin de Rios, M. 1972. *Visionary Vine—Psychedelic Healing in the Peruvian Amazon*. San Francisco: Chandler Publishing Co.

———. 1984. *Hallucinogens: Cross-Cultural Perspectives*. Albuquerque: University of New Mexico Press.

Furst, P. 1976. *Hallucinogens and Culture*. San Francisco: Chandler & Sharp.

———, ed. [1972] 1990. *Flesh of the Gods*. Prospect Heights, IL: Waveland Press.

Grinspoon, L., and J. Bakalar. 1979. *Psychedelic Drugs Reconsidered*. New York: Basic Books.

Grof, S. 1980. *LSD Psychotherapy*. Pomona, CA: Hunter House.

———. 1985. *Beyond the Brain: Birth, Death and Transcendence in Psychotherapy.* Albany, NY: State University of New York Press.

Grunwell, J. 1998. Ayahuasca tourism in South America. *MAPS Bulletin* 8, no. 3.

Harner, M., ed. 1973. *Hallucinogens and Shamanism.* New York: Oxford University Press.

———. 1980. *The Way of the Shaman.* San Francisco: Harper & Row.

Hofmann, A. 1979. *LSD—Mein Sorgenkind.* Stuttgart: Klett-Cotta.

James, W. [1890] 1952. *Principles of Psychology.* Great Books of the Western World. Chicago: Encyclopedia Britannica.

———. [1901] 1958. *Varieties of Religious Experience.* New York: New American Library.

———. [1912] 1996. *Essays in Radical Empiricism.* Lincoln: University of Nebraska Press.

Lamb, F. B. 1971. *Wizard of the Upper Amazon.* Boston: Houghton-Mifflin Co.

———. 1985. *Rio Tigre and Beyond.* Berkeley: North Atlantic Books.

Leary, T., R. Metzner, and R. Alpert. 1964. *The Psychedelic Experience—A Manual Based on the Tibetan Book of the Dead.* New Hyde Park, NY: University Press Books.

Leary, T. [1966] 1997. *Psychedelic Prayers.* Berkeley: Ronin Publishing Co.

Lee, M., and B. Shlain. 1985. *Acid Dreams, The CIS, LSD and the Sixties Rebellion.* New York: Grove Press.

Luna, L. E. 1986. *Vegetalismo—Shamanism Among the Mestizo Population of the Peruvian Amazon.* Stockholm: Almqvist & Wiksell International.

MAPS (Bulletin of the Multidisciplinary Association for Psychedelic Studies). 1998. Vol. 8, no. 3.

McKenna, T., and D. McKenna. 1975. *The Invisible Landscape.* New York: Seabury Press.

McKenna, T. 1993. *True Hallucinations.* New York: Harper Collins.

Metzner, Ralph. 1989. States of consciousness and transpersonal psychology. In *Existential and Phenomenological Perspectives in Psychology,* ed. R. Vallee and S. Halling, 329–38. New York: Plenum Press.

———. 1994. Addiction and transcendence as altered states of consciousness. *Journal of Transpersonal Psychology* 26 (1): 1–17.

———. 1997. *The Unfolding Self—Varieties of Transformative Experience.* Novato, CA: Origin Press.

————. 1998. From Harvard to Zihuatanejo. In *Timothy Leary—Outside Looking In*, ed. R. Forte. Rochester, VT: Inner Traditions International.

————. 1998. Hallucinogenic drugs and plants in psychotherapy and shamanism. *Journal of Psychoactive Drugs* 30 (4):333–41.

Naranjo, C. 1973. *The Healing Journey*. New York: Random House.

Narby, J. 1998. *The Cosmic Serpent—DNA and the Origins of Knowledge*. New York: Jeremy P. Tarcher/ Putnam.

Ott, J. 1993. *Pharmacotheon. Entheogenic Drugs, Their Plant Sources and History*. Kennewick, WA: Natural Products Co.

————. 1994. *Ayahuasca Analogues*. Kennewick, WA: Natural Products Co.

————. 1995. *The Age of Entheogens & The Angels' Dictionary*. Kennewick, WA: Natural Products Co.

Passie, T. 1997. *Psycholytic and Psychedelic Therapy Research—1931–1995*. Hannover, Germany: Laurentius Publishers.

Rätsch, C. 2005. *The Encyclopedia of Psychoactive Plants*, trans. John R. Baker. Rochester, VT: Park Street Press.

Reichel-Dolmatoff, G. 1972. The cultural context of an aboriginal hallucinogen: *Banisteriopsis Caapi*. In *Flesh of the Gods—The Ritual Use of Hallucinogens*, ed. Peter Furst. Prospect Heights, IL: Waveland Press.

Stevens, J. 1987. *Storming Heaven—LSD and the American Dream*. New York: Atlantic Monthly Press.

Taussig, M. 1987. *Shamanism, Colonialism and the Wild Man—A Study in Terror and Healing*. Chicago: University of Chicago Press.

Wasson, R. G. 1980. *The Wondrous Mushroom*. New York: McGraw-Hill.

1

AYAHUASCA:
AN ETHNOPHARMACOLOGIC
HISTORY

DENNIS J. MCKENNA, PH.D.

INTRODUCTION

Of the numerous plant hallucinogens utilized by indigenous populations of the Amazon Basin, perhaps none is as interesting or complex,
botanically, chemically, or ethnographically, as the hallucinogenic beverage known variously as *ayahuasca, caapi,* or *yagé.* The beverage is
most widely known as ayahuasca, a Quechua term meaning "vine of
the souls," which is applied both to the beverage itself and to one of the
source plants used in its preparation, the Malpighiaceous jungle liana,
Banisteriopsis caapi (Schultes 1957). In Brazil, transliteration of this
Quechua word into Portuguese results in the name, hoasca. Ayahuasca,
or *hoasca,* occupies a central position in mestizo ethnomedicine, and
the chemical nature of its active constituents and the manner of its use
make its study relevant to contemporary issues in neuropharmacology,
neurophysiology, and psychiatry.

WHAT IS AYAHUASCA?

In a traditional context, ayahuasca is a beverage prepared by boiling or
soaking the bark and stems of *Banisteriopsis caapi* together with various admixture plants. The admixture employed most commonly is the

Rubiaceous genus *Psychotria*, particularly *P. viridis*. The leaves of *P. viridis* contain alkaloids that are necessary for the psychoactive effect. Ayahuasca is unique in that its pharmacological activity is dependent on a synergistic interaction between the active alkaloids in the plants. One of the components, the bark of *Banisteriopsis caapi*, contains ß-carboline alkaloids, which are potent MAO inhibitors; the other components, the leaves of *Psychotria viridis* or related species, contain the potent short-acting psychoactive agent *N,N*-dimethyltryptamine (DMT). DMT is not orally active when ingested by itself but can be rendered orally active in the presence of a peripheral MAO inhibitor, and this interaction is the basis of the psychotropic action of ayahuasca (McKenna, Towers, and Abbott 1984). There are also reports (Schultes 1972) that other *Psychotria* species are similarly utilized in other parts of the Amazon. In the northwest Amazon, particularly in the Colombian Putumayo and Ecuador, the leaves of *Diplopterys cabrerana*, a jungle liana in the same family as *Banisteriopsis*, are added to the brew in lieu of the leaves of *Psychotria*. The alkaloid present in *Diplopterys*, however, is identical to that in the *Psychotria* admixtures, and pharmacologically, the effect is similar. In Peru, various admixtures in addition to *Psychotria* or *Diplopterys* are frequently added, depending on the magical, medical, or religious purposes for which the drug is being consumed. Although a virtual pharmacopoeia of admixtures are occasionally added, the most commonly employed admixtures (other than *Psychotria*, which is a constant component of the preparation) are various Solanaceous genera, including tobacco (*Nicotiana* sp.), *Brugmansia* sp., and *Brunfelsia* sp. (Schultes 1972; McKenna et al. 1995). These Solanaceous genera are known to contain alkaloids, such as nicotine, scopalamine, and atropine, which affect both central and peripheral adrenergic and cholinergic neurotransmission. The interactions of such agents with serotonergic agonists and MAO inhibitors are essentially unknown in modern medicine.

FOCUS OF THE
PRESENT HISTORICAL PERSPECTIVE

The present chapter presents a brief overview of the history of ethnopharmacological investigations of ayahuasca, which has been a topic of fascination to ethnographers, botanists, chemists, and pharmacologists ever since it first became known to science in the mid-nineteenth century. For expository purposes, the history of ayahuasca ethnopharmacology can be divided into several segments, starting with the prehistoric origins of the beverage and leading up to the present, where ayahuasca is still an active area of research. The modern history of ayahuasca can be dated from the mid-nineteenth century. The focus of the present chapter is on the ethnopharmacologic history of ayahuasca, though it should be noted that this unique beverage has historically impacted religion, politics, and society, as well as science, (e.g., in the Brazilian goverment's acceptance of the legitimacy of the sacramental use of ayahuasca beverages by the UDV and other Brazilian syncretic sects) and the implications and consequences of its continued and spreading use is likely to be felt on a number of levels now and in the future.

PREHISTORIC ROOTS OF AYAHUASCA

The origins of the use of ayahuasca in the Amazon Basin are lost in the mists of prehistory. No one can say for certain where the practice may have originated, and about all that can be stated with certainty is that it was already spread among numerous indigenous tribes throughout the Amazon Basin by the time ayahuasca came to the attention of Western ethnographers in the mid-nineteenth century. This fact alone argues for its antiquity; beyond that, little is known. Plutarco Naranjo, the Equatorian ethnograper, has summarized what little information is available on the prehistory of ayahuasca (Naranjo 1979, 1986). There is abundant archeological evidence, in the form of pottery vessels, anthropomorphic figurines, snuffing trays and tubes, etc., that plant hallucinogen use was well established in the Ecuadorian Amazon by 1500–2000 B.C. Unfortunately, most of the specific evidence, in the form of vegetable powders, snuff trays, and pipes, is related to the use of psychoactive plants other than ayahuasca, such as coca, tobacco, and the hallucinogenic snuff derived from *Anadenanthera* species and known as vilka and various other names. There is nothing in the form of iconographic materials or preserved botanical remains that would unequivocally establish the prehistoric use of ayahuasca, although it is probable that these pre-Colombian cultures, sophisticated as they were in the use of a variety of psychotropic plants, were also familiar with ayahuasca and its preparation. The lack of data is frustrating, however, particularly in respect to a question that has fascinated ethnopharmacologists since the late 1960s when its importance was first brought to light through the work of Richard Schultes and his students. As mentioned above, ayahuasca is unique among plant hallucinogens in that it is prepared from a combination of two plants: the bark or stems of *Banisteriopsis* species, together with the leaves of *Psychotria* species or other DMT-containing admixtures. The beverage depends on this unique combination for its activity. There seems small likelihood of accidentally combining the two plants to obtain an active preparation when neither is particularly active alone, yet we know that at some point in prehistory, this fortuitous combination was discovered. At that point, ayahuasca was "invented." Just how this discovery was made, and who was responsible, we may never know, though there are several charming myths that address the topic. Mestizo ayahuasqueros in Peru will, to this day, tell you that this knowledge comes directly from the "plant teachers" (Luna 1984), while the mestres of the Brazilian syncretic cult, the UDV, will tell you with equal conviction that the knowledge came from "the first scientist," King Solomon, who imparted the technology to the Inca king during a little publicized visit to the New World in antiquity. In the absence of data, these explanations are all that we have. All that we can say with confidence is that the knowledge of the techniques for preparing ayahuasca, including knowledge of the appropriate admixture plants, had diffused throughout the Amazon by the time the use of ayahuasca came to the attention of any modern researcher.

SCIENTIFIC DISCOVERY OF AYAHUASCA—
THE NINETEENTH CENTURY

The archeological prehistory of ayahuasca is likely to remain inextricably bound up with its mythical origins for the rest of time, unless some artifact should be uncovered that would unequivocally establish the antiquity of its usage.

By contrast, what might be called the modern or the scientific history of ayahuasca is traceable to 1851, when the great English botanist Richard Spruce encountered the use of an intoxicating beverage among the Tukano Indians of the Rio Uapes in Brazil (Schultes 1982). Spruce collected flowering specimens from the large jungle liana used as the source of the beverage, and this collection was the basis for his classification of the plant as *Banisteria caapi*; it was reclassified as *Banisteriopsis caapi* by the taxonomist Morton in 1931 as part of his revision of the generic concepts within the family Malpighiaceae.

Seven years later, Spruce again encountered the same liana in use among the Guahibo Indians on the upper Orinoco of Colombia and Venezuela, and, later the same year, found the Záparo Indians of Andean Peru taking a narcotic beverage, prepared from the same plant, which they called ayahuasca. Although Spruce's discovery predates any other published accounts, he did not publish his findings until 1873, when it was mentioned in a popular account of his Amazon explorations (Spruce 1873). A fuller exposition was not to appear until Spruce published his account in A. R. Wallace's anthology in 1908, *Notes of a Botanist on the Amazon and Andes* (Spruce 1908). Credit for the earliest published reports of ayahuasca usage belongs to the Ecuadorian geographer Manuel Villavicencio, who, in 1858, wrote of the use of ayahuasca in sorcery and divination on the upper Rio Napo (Villavicencio 1858). Although Villavicencio supplied no botanical details about the plant used as the source of the beverage, his account of his own self-intoxication left no doubt in Spruce's mind that they were writing about the same thing.

Throughout the remainder of the nineteenth century, various ethnographers and explorers continued to report on their encounters of the use of an intoxicating beverage prepared by various indigenous Amazonian tribes, and purportedly prepared from the "roots" (Crévaux 1883), of various "shrubs" (Koch-Grünberg 1909) or "lianas" (Rivet 1905) of uncertain botanical provenance. Unlike Spruce, who had the presence of mind, not only to collect botanical voucher specimens, but also materials designated for eventual chemical analysis, these later investigators did not collect specimens of the plants they observed, and hence their accounts are now of little more than historical importance. One notable exception was Simson's (1886) publication of the use of ayahuasca among Ecuadorian Indians, noting that they "drank ayahuasca mixed with yage, sameruja leaves, and guanto wood, an indulgence which usually results in a broil between at least the partakers of the beverage." None of the ingredients were identified, nor were voucher specimens collected, but this report is the earliest indication that other admixture species were employed in the preparation of ayahuasca.

While Richard Spruce and other adventurous Amazonian explorers were collecting the first field reports of ayahuasca from 1851 onward, the groundwork was already being laid for important work on the chemistry of ayahuasca that would take place in the second decade of the twentieth century. The nineteenth century witnessed the birth of natural products chemistry, starting with the isolation of morphine from opium poppies by the German pharmacist Sertüner in 1803. A disproportionate number of natural products isolated for the first time during this period were alkaloids, probably because these bases are relatively easy to isolate in a pure form, and partly because the plants that contain them were and are important drug plants with obvious and often dramatic pharmacological properties. It was during this period of feverish alkaloid discovery that German chemist H. Göbel isolated harmaline from the seeds of the Syrian rue, *Peganum harmala*. Six years later, his colleague J. Fritsch isolated harmine from the seeds in 1847. More than fifty years later, a third alkaloid, harmalol, was also isolated from Syrian rue seeds by Fisher in 1901. Harmine, like the other ß-carbolines named after the species epithet of *Peganum harmala*, would later turn out to be identical to the major ß-carboline found in *Banisteriopsis caapi*; the definitive establishment of the equivalence of the ayahuasca ß-carboline to harmine from Syrian rue, however, would not take place until the 1920s, after harmine had been independently isolated by several investigators and given a variety of names. The final nineteenth-century event of significance in the scientific history of ayahuasca took place in 1895, with the first investigations of the effects of harmine on the central nervous system in lab animals by Tappeiner; his preliminary results were followed up more systematically by Gunn in 1909, who reported that the major effects were motor stimulation of the central nervous system with tremors and convulsions, followed or accompanied by paresis and slowed pulse (Gunn 1935).

AYAHUASCA IN THE EARLY TWENTIETH CENTURY
(1900–1950)

The early decades of the twentieth century witnessed the publication of Spruce's detailed accounts of his Amazonian explorations and his observations of the use of the narcotic beverage among several tribes that he contacted. Although brief reports had been published earlier by Spruce and others, it was Spruce's account of his travels in a volume edited by the famed naturalist and codiscoverer of evolution A. R. Wallace in 1908 that may have rescued the knowledge of ayahuasca from the depths of academic obscurity and brought it to the attention of educated lay people.

During this early twentieth-century period, progress in the understanding of ayahuasca took place mainly on two fronts: taxonomic, and chemical. With some notable exceptions, pharmacological investigations of the properties of ayahuasca were relatively quiescent during this period.

The botanical history of ayahuasca during this period is an amusing combination of excellent taxonomic detective work by some, and egregious errors compounded upon errors by others. Safford, in 1917, asserted his belief that ayahuasca and the beverage known as caapi were identical and derived from the same plant. The French anthropologist Reinberg (1921) compounded the confusion by his assertion that ayahuasca was referable to *Banisteriopsis caapi*, but that yagé was prepared from an Apocyanaceous genus, *Haemadictyon amazonicum*, now correctly classified as *Prestonia amazonica*. This error, which apparently originated from an uncritical reading of Spruce's original field notes, was to persist and propagate through the literature on ayahuasca for the next forty years. It was finally put to rest when Schultes and Raffauf published a paper specifically refuting this misidentification (Schultes and Raffauf 1960), however, it still crops up occasionally in technical literature.

Among the investigators who helped to clarify, rather than cloud, the taxonomic understanding of ayahuasca botany must be mentioned the works of Rusby and White in Bolivia in 1922 (White 1922) and the publication by Morton in 1930 of the field notes made by the botanist Klug in the Colombian Putumayo. From Klug's collections, Morton described a new species of *Banisteriopsis*, *B. inebriens*, used as a hallucinogen, but he also asserted that at least three species, *B. caapi*, *B. inebriens*, and *B. quitensis*, were used similarly and that two other species, *Banisteria longialata*, and *Banisteriopsis rusbyana* may have been used as admixtures to the preparation. Curiously, it was two chemists, Chen and Chen (1939), who did the most to clarify the early taxonomic confusion about the identity of the ayahuasca source plants. These investigators, working on the isolation of the active principles of yagé and ayahuasca, supported their investigations with authentic botanical voucher specimens (a rare practice at that time) and, after a review of the literature, concluded that caapi, yagé, and ayahuasca were all different names for the same beverage, and that their source plant was identical: *Banisteriopsis caapi*. Subsequent work by Schultes and others in the 1950s would establish that, in fact, Malpighiaceous species other than *B. caapi* were implicated in the preparation of the beverage, but considering the reigning confusion of the time, Chen and Chen's contribution was a rare light in the forest of prevailing darkness. From subsequent fieldwork, it is now quite clear that the two main botanical sources of the beverage variously known as caapi, ayahuasca, yagé, natéma, and pinde are the barks of *B. caapi* and *B. inebriens*.

The first half of the twentieth century was also the period in which the first serious chemical investigations of the active principles of ayahuasca were carried out. Like much of the initial taxonomic work taking place during this same period, scientific progress on this front was marked at first by confusion arising from the simultaneous investigations of several independent groups of investigators. Gradually, as these investigations found their way into the scientific literature, clarity began to emerge from a fairly murky picture.

Harmine, which consensus would eventually establish as the major ß-carboline alkaloid of *Banisteriopsis* species, had been isolated from the seeds of *Peganum harmala* in 1847 by the German chemist Fritsch. Its unequivocal identification was still several decades in the future when an alkaloid named "telepathine" was obtained from unvouchered botanical material called "yajé" by Zerda and Bayón in 1905 (quoted in Perrot and Hamet 1927). In 1923, an alkaloid was again isolated from unvouchered botanical materials by the Colombian chemist Fisher

Cardenas (1923) and was also named telepathine; at the same time, another Colombian team, chemists Barriga-Villalba and Albarracin (1925) isolated an alkaloid, yageine. This may also have been harmine in an impure form, but the formula assigned at the time and the melting point were inconsistent for a ß-carboline structure. To compound the confusion, the vine with which Barriga-Villalba worked had been "identified" as *Prestonia amazonica*, but he later revised this identification to *Banisteriopsis caapi*. In all of these instances, the lack of botanical reference specimens rendered the work of dubious value.

Things began to get slightly better from 1926 into the 1950s. Michaels and Clinquart (1926) isolated an alkaloid that they called yageine from unvouchered materials. Shortly afterward, Perrot and Hamet (1927) isolated a substance that they called telepathine and suggested that it was identical to yageine. Lewin, in 1928, isolated an alkaloid that he named banisterine; this was shown to be identical with harmine, previously known from the Syrian rue, by chemists from E. Merck and Co. (Elger 1928; Wolfes and Rumpf 1928). Elger worked from vouchered botanical materials that had been identified at Kew Gardens as *Banisteriopsis caapi*. At Lewin's urging, based on his own animal studies, the pharmacologist Kurt Beringer (1928) used samples of "banisterine" donated by Lewin in a clinical study of fifteen post-encephalitic Parkinson's patients and reported dramatic positive effects (Beringer 1928). This was the first time that a reversible MAO inhibitor had been evaluated for the treatment of Parkinson's disease, though harmine's activity as a reversible MAOI was not discovered until nearly thirty years later. It also represents one of the few instances where a hallucinogenic drug has been clinically evaluated for the treatment of any disease (Sanchez-Ramos 1991).

Working from vouchered botanical materials supplied by Llewellyn Williams of the Chicago Field Museum, Chen and Chen (1939) succeeded in confirming the work of Elger and Wolfes and Rumpf; these workers isolated harmine from the stems, leaves, and roots of *B. caapi* and confirmed its identity with banisterine, previously isolated by Lewin. In 1957 Hochstein and Paradies analyzed vouchered material of ayahuasca collected in Peru and isolated harmine, harmaline, and tetrahydroharmine. The investigations of the constituents of other

Banisteriopsis species was not undertaken until 1953, when O'Connell and Lynn (1953) confirmed the presence of harmine in the stems and leaves of vouchered specimens of *B. inebriens* supplied by Schultes. Subsequently Poisson (1965) confirmed these results by isolating harmine and a small amount of harmaline from "natema" from Peru, identified by Cuatrecasas as *B. inebriens*.

MID-TWENTIETH CENTURY (1950–1980)

The first half of the twentieth century witnessed the initial scientific studies of ayahuasca and began to shed some light on the botanical sources of this curious hallucinogen and the nature of its active constituents. During the three decades from 1950 to 1980, botanical and chemical studies continued apace, and new discoveries laid the groundwork for an eventual explanation of the unique pharmacological actions of ayahuasca.

On the chemical front, the work of Hochstein and Paradies (1957) confirmed and extended the previous work of Chen and Chen (1939) and others. The active alkaloids of *Banisteriopsis caapi* and related species were now firmly established as harmine, tetrahydroharmine, and harmaline. In the late 1960s however, the first detailed reports of the use of admixtures as a regular, if not invariant, component of the ayahuasca brew began to emerge (Pinkley 1969), and it soon became apparent that at least two of these admixtures, *Banisteriopsis rusbyana* (later reclassified by Bronwen Gates as *Diplopterys cabrerana*) and *Psychotria* species, particularly *P. viridis*, (Schultes 1967) were added to the brew to "strengthen and extend" the visions. A further surprise came when the alkaloid fractions obtained from these species proved to be the potent short-acting (but orally inactive) hallucinogen *N,N*-dimethyltryptamine (DMT) (Der Marderosian et al. 1968). This compound had been known as a synthetic for some decades following Manske's initial synthesis; but its occurrence in nature and its hallucinogenic properties had only come to light a few years earlier, when Fish, Johnson, and Horning (1955) had isolated it as the putative active principle in *Piptadenia peregrina* (later reclassified as *Anadenanthera peregrina*), the source of a hallucinogenic snuff used by Indians of the Carribean, as well as the Orinoco basin of South America.

The pharmacological rationale for the discovery by Schultes, Pinkley, and others in the late 1960s that ayahuasca depended for its activity on a synergistic interaction between the MAO-inhibiting ß-carbolines in *Banisteriopsis* with the psychoactive but peripherally inactivated tryptamine DMT had already been provided in 1958 by Udenfriend and coworkers (Udenfriend et al. 1958). These researchers in the Laboratory of Clinical Pharmacology at NIH were the first to demonstrate that ß-carbolines were potent, reversible inhibitors of MAO. During this same period, clinical work and self-experimentation by the Hungarian psychiatrist and pharmacologist Stephen Szara (1957) with the newly synthesized DMT lead to the publication of the first reports of its profound, though short-lasting, hallucinogenic actions in humans. Szara's experiments also lead to the first recognition that the compound is not orally active, though the mechanisms of its inactivation on oral administration were not fully understood. Ironically, several decades later, the DMT pioneer Szara would be appointed as the head of NIDA (National Institute on Drug Abuse).

In 1967, during the height of the Summer of Love in Haight-Ashbury, a unique symposium was held in San Francisco under the sponsorship of what was at the time the U.S. Department of Health, Education, and Welfare. Entitled *Ethnopharmacologic Search for Psychoactive Drugs* (the proceedings were later published under that title as U.S. Public Health Service Publication No. 1645, issued by the U.S. Government Printing Office) (Efron et al. 1967) this conference brought together the leading lights of the day in the emerging field of psychedelic ethnopharmacology. Participants included toxicologist Bo Holmstedt of the Karolinska Institute in Stockholm, ethnobotanist Richard Evans Schultes, chemist Alexander Shulgin, newly credentialed M.D. and marijuana researcher Andrew Weil, and others. It was the first time that a conference on the botany, chemistry, and pharmacology of psychedelics had been held, and as it happened, it was certainly the last time that such a conference would be held under government sponsorship. This landmark conference, and the publication issuing from it, which was to become a classic of psychedelic literature, was the first forum where the state of the art at the time regarding ayahuasca in its multidisciplinary aspects was revealed to the world. The symposium volume included chapters on the chemistry of ayahuasca (Deulofeu 1967), the ethnography of its use and preparation (Taylor 1967), and the human psychopharmacology of the ß-carbolines of ayahuasca (Naranjo 1967). It is an ironic commentary on the paucity of knowledge of ayahuasca at the time that the uses of tryptamine-containing admixtures, and their activation via MAO-inhibition, did not even surface for discussion at the symposium; the prevailing assumption was that the psychoactivity of ayahuasca was due primarily if not entirely to the ß-carbolines.

In the five years following this conference, progress was made in understanding ayahausca pharmacology and chemistry. Schultes and his students Pinkley and der Marderosian published their initial findings on the DMT-containing admixture plants (Der Marderosian et al. 1968; Pinkley 1969), fueling speculation that DMT, orally activated by ß-carbolines, was responsible for much of the activity of the brew. This notion, although plausible, would not be scientifically confirmed for another decade.

In 1972, Rivier and Lindgren (1972) published one of the first interdisciplinary papers on ayahuasca, reporting on the alkaloid profiles of ayahuasca brews and source plants collected among the Shuar people of the upper Rio Purús in Peru. At the time, their paper was one of the most thorough chemical investigations of the composition of ayahuasca brews and source plants that referenced vouchered botanical collections. It also discussed numerous admixture plants other than the *Psychotria* species and *Diplopterys cabrerana,* and for the first time provided evidence indicating that ayahuasca admixture technology was complex, and that many species were on occasion used as admixtures.

In the later 1970s a team of Japanese phytochemists became interested in the chemistry of *Banisteriopsis* and reported the isolation of a number of new ß-carbolines and the pyrrolidine alkaloids shihunine and dihydroshihunine (Hashimoto and Kawanishi 1975, 1976; Kawanishi et al. 1982). Most of the newly reported ß-carbolines were isolated in extreme trace amounts, however, and the possibility was later raised that they might be artifacts resulting from the isolation procedures (McKenna et al. 1984).

LATE TWENTIETH CENTURY (1980–PRESENT)

Following publication of Rivier and Lindgren's paper, there was little further progress on the scientific front for the remainder of the 1970s. There was no comparable follow-up to Rivier and Lindgren's work until Terence McKenna et al. (1984) published the results of their chemical, ethnobotanical, and pharmacological investigations of ayahuasca and its admixtures, based on vouchered botanical specimens and samples of brews used by mestizo ayahuasqueros in Peru. This paper was significant because it represented the first time that the theory proposed to explain the oral activity of the beverage was experimentally confirmed. The active principal was shown to be DMT, rendered orally active by ß-carboline-mediated blockade of peripheral MAO. Assays of ayahuasca fractions in rat-liver MAO systems showed that the brews were extremely potent MAO inhibitors even when diluted many orders of magnitude. A further important discovery was the finding that the levels of alkaloids typically found in the mestizo ayahuasca brews exceeded the levels found in the upper Rio Purús ayahuasca by Rivier and Lindgren, sometimes by an order of a magnitude or more. Based on the known human pharmacology of DMT and ß-carbolines, Terence McKenna and coworkers showed that a typical dose (100 ml) of the mestizo ayahuasca samples contained enough DMT to constitute an active dose. The investigators suggested that the lower levels of alkaloids found in the Shuar samples of Rivier and Lindgren (1972) may have resulted from the different methods used in preparation. The Shuar typically soak the *Banisteriopsis* and admixture plants in cold water; they do not boil the plants, nor do they reduce the volume of the final extract, as is typically done in mestizo practice. These factors explained the discrepancies in alkaloid concentration found in the two different studies or at least provided a plausible rationale to explain the differences.

The decade of the 1980s also witnessed the early contributions of the anthropologist, Luis Eduardo Luna. Working among mestizo ayahuasqueros near the cities of Iquitos and Pucallpa in Peru, Luna's work was the first to articulate the importance of the strict diet followed by apprentice shamans, as well as the specific uses of some of the more unusual admixture plants (Luna 1984a; 1984b; 1986). He was also the first to report on the concept of "plant teachers" *(plantas que enseñan)*, which is how many of the admixture plants are viewed by the mestizo ayahuasqueros. In 1986, McKenna, Luna, and Towers published the first comprehensive tabulation of the species used as admixtures and the biodynamic constituents contained in them, pointing out that these relatively uninvestigated species comprise an extensive folk pharmacopoeia worthy of closer scrutiny as potential sources of new therapeutic agents (McKenna et al. 1995).

While conducting fieldwork together in the Peruvian Amazon in 1985, McKenna and Luna first began discussing the possibility of conducting a biomedical investigation of ayahuasca. The superior health of the ayahuasqueros, even at advanced ages, seemed remarkable and something that could be amenable to scientific study. The logistical challenges of carrying out such work in Peru, however, seemed daunting, since access to storage facilities for plasma samples was limited and local concepts of witchcraft made it unlikely that ayahuasqueros would submit to medical procedures such as collection of blood and urine samples. The workers wrote a preliminary proposal for the project following their return from the field but did not pursue funding.

In 1991, however, a fresh opportunity to initiate such a study presented itself in Brazil. McKenna and Luna were among several foreigners invited to participate in a conference in São Paulo by the Medical Studies section of the União do Vegetal (UDV), a Brazilian syncretic religion that used ayahuasca in their ceremonies. The group's use of ayahuasca in a ritual context (under the names *hoasca*, *vegetal*, or simply *cha*, "tea"), while permitted by the Brazilian regulatory authorities, was subject to provisional review. Many members of the UDV were themselves physicians, psychiatrists, or had other kinds of medical expertise and so were most receptive to the notion of conducting a biomedical study of ayahuasca when it was proposed to them by Luna and McKenna. It turns out that this had been a part of their own unspoken agenda all along and was part of the reason for inviting the foreign investigators to the first Medical Studies Conference on Hoasca. Besides the opportunity to satisfy scientific curiosity about the human pharmacology of hoasca, the UDV had a political motive for carrying out such a study; they wanted to be able to demonstrate to the Brazilian health authorities that the long-term use of hoasca tea was safe, and did not cause addiction or other adverse reactions. The UDV physicians were hoping to enlist foreign scientists to collaborate in the study. The question of how the study was to be funded had yet to be answered.

Following the 1991 conference, McKenna returned to the United States and drafted a proposal describing the objectives of the study that was to become known as the Hoasca Project. Initially, the objective was to submit the proposal to the National Institute on Drug Abuse, but as the proposal took shape it became clear that funding for the study would be unlikely to originate from any government agency. Not only were there legal, logistical, and political problems with securing NIH funds for a study to be carried out in Brazil, it was also clear that given the nature of government drug policy, the NIH would not look favorably on a proposal that was not aimed at demonstrating serious harmful consequences resulting from the use of a psychedelic drug. Fortunately, McKenna had affiliations with Botanical Dimensions, a nonprofit organization dedicated to the investigation of ethnomedically important plants, and through this venue he was able to solicit generous grants from several private individuals.

With sufficient funding assured for at least a modest pilot study, McKenna enlisted the collaborative talents of various colleagues in the medical and academic communities. Eventually, a truly international, interdisciplinary study team was formed, consisting of scientists from UCLA, the University of Miami, the University of Kuopio in Finland, the University of Rio de Janeiro, University of Campinas near São Paulo, and the Hospital Amazonico in Manaus.

The team returned to Manaus in the summer of 1993 to begin the field phase of the research, which was conducted using volunteers who were members of the Nucleo Caupari in Manaus, one of the oldest and largest UDV congregations in Brazil. The team spent five weeks in Brazil administering test doses of hoasca tea to the volunteers, collecting plasma and urine samples for later analysis, and carrying out a variety of physiological and psychological measurements.

The result was one of the most comprehensive multifaceted investigations of the chemistry, psychological effects, and psychopharmacology of a psychedelic drug to be carried out in this century. Both the

Three stages in the preparation
of the hoasca tea, by members
of the UDV in Brazil
(photos by Leonide D. Principe)

acute and the long-term effects of regular ingestion of hoasca tea were measured and characterized; extensive psychological evaluations, and in-depth structured psychiatric interviews were conducted with all volunteers; the nature of the serotonergic response to ayahuasca was measured and characterized; and the pharmacokinetics of the major hoasca alkaloids were measured for the first time in human plasma. Since completion of the field phase of the study, the results have been published in a number of peer-reviewed papers (Grob et al. 1996; Callaway et al. 1994, 1996, 1998) and have recently been summarized in a comprehensive review (McKenna et al. 1998). Among the key findings were that long-time members of the UDV commonly underwent experiences that changed their lives and behavior in positive and profound ways; and that there was a persistent elevation in serotonin uptake receptors in platelets, possibly indicative of similar long-term serotonergic modulation occurring in the central nervous system that may reflect long-term adaptive changes in brain functions. The study did establish that the regular use of hoasca, at least within the ritual context and supportive social environment that exists within the UDV, is safe and without adverse long-term toxicity, and, moreover, apparently has lasting, positive influences on physical and mental health.

THE FUTURE OF AYAHUASCA RESEARCH

The field and laboratory phases of the Hoasca Project have been completed for sometime, and now that the last and final major paper resulting from the work has been accepted for publication, the Project is in its final stages. Always conceived as a pilot study, the objectives of the hoasca study were modest and intended to indicate directions for future research. In this regard, the study was a remarkable success; like all good science, the study raised more questions than it answered and suggested several promising directions for future research. Now that ayahuasca has been clearly shown to be safe, nontoxic, and therapeutically useful as medicine, it is to be hoped that future researchers will devote sufficient interest, as well as funds, to the exploration of its healing potential.

SOME SPECULATIVE ISSUES

With the completion of the Hoasca Project, there now exists a solid foundation of basic data to serve as the underpinning of future scientific investigations as their focus moves from the field to the laboratory and the clinic. But outside the perimeter of the cold light of reason cast by scientific scrutiny, there remain a number of issues surrounding ayahuasca that are unlikely be resolved by science alone, at least not by scientific methods as they are now understood. Ayahuasca is a symbiotic ally of the human species; its association with our species can be traced at least as far back as New World prehistory. The lessons we have acquired from it, in the course of millennia of coevolution, may have profound implications for what it is to be human, and to be an intelligent, questioning species within the biospheric community of species.

Although we have no certain answers, the question of the nature and meaning of the relationship between humanity and this visionary vine, and by extension with the entire universe of plant teachers, persistently troubles us. Why should plants contain alkaloids that are close analogs of our own neurotransmitters, and that enable them to "talk" to us? What "message" are they trying to convey, if any? Was it purely happenstance, purely accident, that led some early, experiment-minded shaman to combine the ayahuasca vine and the chacruna leaf, to make the tea that raised the curtain on the "invisible landscape" for the first time? It seems unlikely, since neither of the key ingredients are particulary inviting as food, and yet what else could it have been? The ayahuasqueros themselves will simply tell you that "the vine calls." Others, trying to be more sophisticated and rational, but proffering no more satisfying explanation, will talk about plant alkaloids as interspecies pheromonal messengers and as carriers of sensoritropic cues that enabled early humans to select and utilize the biodynamic plants in their environment.

Still others, such as my brother Terence McKenna and I in our early work, and a more recent reformulation of a similar theory by anthropologist Jeremy Narby (McKenna and McKenna 1975; Narby 1998), argue that by some as yet obscure mechanism, the visionary experiences afforded by plants such as ayahuasca give us an insight—an intuitive understanding—of the molecular bedrock of biological being, and that this intuitive knowledge, only now being revealed to the scientific worldview by the crude methods of molecular biology, has always been available as direct experience to shamans and seers with the courage to forge symbiotic bonds with our mute but infinitely older and wiser plant allies.

Such notions are surely speculative and are certainly not science; but to an observer of the contemporary world, who has been involved both scientifically and personally with ayahuasca for many years now, I find it very interesting that such "wild" speculations keep reasserting themselves, no matter how much we try to desacralize the tea and render it down to a matter of chemistry and botany, receptor sites and pharmacology. All of those things are important, certainly; but none of them will ever explain the undeniable and profound mystery that is ayahuasca.

References

Barriga-Villalba, A. M. 1925. Yajeine. A new alkaloid. *Journal of the Society of Chemistry and Industry* 44:205–207.

Beringer, K. 1928. Über ein neues, auf das extrapyramidal-motorische System wirkendes Alkaloid (Banisterin). *Nervenarzt* 1:265–75.

Callaway, J. C., D. J. McKenna, C. S. Grob, G. S. Brito, L. P. Raymon, R. E. Poland, E. N. Andrade, E. O. Andrade, and D. C. Mash. 1999. Pharmacokinetics of *Hoasca* alkaloids in healthy humans. *Journal of Ethnopharmacology* 65 (3): 243–56.

Callaway, J. C., L. P. Raymon, W. L. Hearn, D. J. McKenna, C. S. Grob, G. S. Brito, and D. C. Mash. 1996. Quantitation of N,N-dimethyltryptamine and harmala alkaloids in human plasma after oral dosing with Ayahuasca. *Journal of Analytical Toxicology* 20:492–97.

Callaway, J. C., M. M. Airaksinen, D. J. McKenna, G. S. Brito, and C. S. Grob. 1994. Platelet serotonin uptake sites increased in drinkers of ayahuasca. *Psychopharmacology* 116:385–87.

Chen, A. L., and K. K. Chen. 1939. Harmine: The alkaloid of caapi. *Quarterly Journal of Pharmacy and Pharmacology* 12:30–38.

Crévaux, J. 1883. *Voyages dans l'Amerique du Sud*. Paris: Librairie hachette & Cie.

Der Marderosian, A. H., H. V. Pinkley, and M. F. Dobbins. 1968. Native use and occurrence of N,N-dimethyltryptamine in the leaves of Banisteriopsis rusbyana. *American Journal of Pharmacy* 140:137.

Deulofeu, V. (1967). Chemical compounds isolated from Banisteriopsis and related species. In *Ethnopharmacological Search for Psychoactive Drugs,* ed. D. H. Efron, B. Holmstedt, and N. S. Kline. U.S. Public Health Service Publication No. 1645. Washington, D.C.: GPO.

Efron, D. H., B. Holmstedt, and N. S. Kline, eds. 1967. *Ethnopharmacological Search for Psychoactive Drugs.* U.S. Public Health Service Publication No. 1645. Washington, D.C.: GPO.

Elger, F. 1928. Über das Vorkommen von Harmin in einer südamerikanischen Liane (Yagé). *Helvetica Chemica Acta* 11:162.

Fischer, C. G. 1923. Éstudio sobre el principio activo de Yagé. Unpublished thesis. Bogotá: Université Nacional.

Fish, M. S., N. M. Johnson, and E. C. Horning. 1955. Piptadenia alkaloids. Indole bases of *Piptadenia peregrina* (L.) Benth and related species. *Journal of the American Chemical Society* 77:5892–95.

Grob, C. S., D. J. McKenna, G. S. Brito, E. S. Neves, G. Oberlender, O. L. Saide, E. Labigalini, C. Tacla, C. T. Miranda, R. J. Strassman, and K. B. Boone. 1996. Human psychopharmacology of hoasca, a plant hallucinogen used in ritual context in Brasil. *Journal of Nervous & Mental Disease* 184:86–94.

Gunn, J. A. 1935. Relationship between chemical constitution, pharmacological actions, and therapeutic uses in the harmine group of alkaloids. *Archives Internationales de Pharmacodynamie* 50:379–96.

Hashimoto, Y., and K. Kawanishi. 1975. New organic bases from the Amazonian *Banisteriopsis caapi. Phytochemistry* 14:1633–35.

———. 1976. New alkaloids from *Banisteriopsis caapi. Phytochemistry* 15:1559–60.

Hochstein, F. A., and A. M. Paradies. 1957. Alkaloids of *Banisteriopsis caapi* and *Prestonia amazonica. Journal of the American Chemical Society* 79:5735ff.

Kawanishi, K., Y. Uhara, and Y. Hashimoto. 1982. Shihunine and dehydroshihunine from *Banisteriopsis caapi. Journal of Natural Products* 45:637–38.

Koch-Grünberg, T. 1909. *Zwei Jahre unter den Indianern* 1:298ff.

Lewin, L. 1928. Sur une substance envirante, la banisterine, extraite de *Banisteria caapi* Spr. *Comptes Rendeus* 186:469ff.

Luna, L. E. 1984a. The healing practices of a Peruvian shaman. *Journal of Ethnopharmacology* 11:123–33.

———. 1984b. The concept of plants as teachers among four mestizo shamans of Iquitos, northeast Peru. *Journal of Ethnopharmacology* 11:135–56.

———. 1986. *Vegitalismo: Shamanism Among the Mestizo Population of the Peruvian Amazon.* Stockholm: Almqvist and Wiksell International.

McKenna, D., G. H. N. Towers, and F. S. Abbott. 1984. Monoamine oxidase inhibitors in South American hallucinogenic plants: Tryptamine and ß-carboline constituents of ayahausca. *Journal of Ethnopharmacology* 10:195–223.

McKenna, D. J., C. S. Grob, and J. C. Callaway. 1998. The scientific investigation of Ayahuasca: A review of past and current research. *Heffter Review of Psychedelic Research* 1:65–77.

McKenna, D. J., L. E. Luna, and G. H. N. Towers. 1995. Biodynamic constituents in Ayahuasca admixture plants: an uninvestigated folk pharmacopoeia. In *Ethnobotany: Evolution of a Discipline,* ed. S. von Reis and R. E. Schultes. Portland: Dioscorides Press.

McKenna, D. J., and T. K. McKenna. 1975. *The Invisible Landscape.* New York: Seabury Press.

Michaels, M., and E. Clinquart. 1926. Sur des réactions chemiques d'identificacion de la yageine. *Bulletin de Academie Royale Médicin Belgique,* Series 5, 6:79ff.

Morton, C. V. 1930. Notes on yagé, a drug plant of southeastern Colombia. *Journal of the Washington Academy of Science* 21:485.

Naranjo, C. 1967. Psychotropic properties of the harmala alkaloids. In *Ethnopharmacological Search for Psychoactive Drugs,* eds. D. H. Efron, B. Holmstedt, and N. S. Kline. U.S. Public Health Service Publication No. 1645. Washington, D.C.: GPO.

Naranjo, P. 1979. Hallucinogenic plant use and related indigenous belief systems in the Ecuadorian Amazon. *Journal of Ethnopharmacology* 1:121–45.

———. 1986. El ayahuasca in la arqueología ecuatoriana. *America Indigena* 46:117–28.

Narby, J. 1998. *The Cosmic Serpent: DNA and the Origins of Knowledge.* New York: Jeremy Tracher/Putnam Publishers.

O'Connell, F. D., and E. V. Lynn. 1953. The alkaloids of *Banisteriopsis inebriens* Morton. *Journal of the American Pharmaceutical Association* 42:753.

Perrot, E., and R. Hamlet. 1927. Le yagé, plante sensorielle des Colombie. *Comptes Rendues de la Adacemie Scientifique* 184:1266.

Pinkley, H. V. 1969. Plant admixtures to ayahuasca, the South American hallucinogenic drink. *Lloydia* 32:305ff.

Poisson, J. 1965. Note sur le "natem," boisson toxique péruvienne. *Annales Pharmacia Francaise* 23:241ff.

Reinberg, P. 1921. Contribution à l'étude des boissons toxiques des indiens du Nord-ouest de l'Amazon, l'ayahuasca, le yagé, le huanto. *Journal de la Societé des Americanistes, Paris,* 4:49ff.

Rivet, P. 1905. Les indiens colorados. *Journal de la Societé des Americanistes, Paris,* 2:201ff.

Rivier, L., and J. Lindgren. 1972. *Ayahuasca,* the South American hallucinogenic drink: Ethnobotanical and chemical investigations. *Economic Botany* 29:101–129.

Sanchez-Ramos, J. R. 1991. Banisterine and Parkinson's Disease. *Clinical Neuropharmacology* 14:391–402.

Schultes, R. E. 1967. The place of ethnobotany in the ethnopharmacologic search for psychoactive drugs. In *Ethnopharmacologic Search for Psychoactive Drugs,* ed. D. H. Efron, B. Holmstedt, and N. S Kline. U.S. Public Health Service Publication No. 1645. Washington, D.C.: GPO.

———. 1982. The beta-Carboline hallucinogens of South America. *Journal of Psychoactive Drugs* 14:205–19.

Schultes, R. E., and R. Raffauf. 1960. Prestonia: An Amazonian narcotic or not? *Botanical Museum Leaflets,* Harvard University 19:109–22.

Simson, A. 1886. *Travels in the Wilds of Ecuador.* London: Lowe, Livinston, Marston & Searle.

Spruce, R. A. 1873. On some remarkable narcotics of the Amazon Valley and Orinoco. Ocean Highways. *Geographical Magazine* 1:184–193.

———. 1908. *Notes of a Botanist on the Amazon and Andes,* ed. A. R. Wallace, 2 vols. London: MacMillan.

Szára, S. I. 1957. The comparison of the psychotic effects of tryptamine derivatives with the effects of mescaline and LSD-25 in self-experiments. In *Psychotropic Drugs,* ed. S. Garratini and V. Ghetti, 460–67. New York: Elsevier.

Taylor, D. 1967. The making of the hallucinogenic drink from *Banisteriopsis caapi* in northern Peru. In *Ethnopharmacological Search for Psychoactive Drugs,* ed. D. H. Efron, B. Holmstedt, and N. S. Kline. U.S. Public Health Service Publication No. 1645. Washington, D.C.: GPO.

Udenfriend, S., B. Witkop, B. G. Redfield, and H. Weissbach. 1958. Studies with reversible inhibitors of monoamine oxidase: Harmaline and related compounds. *Biochemical Pharmacology* 1:160–65.

Villavicencio, M. 1858. *Geografía de la República del Ecuador.* New York: Craighead.

White, O. E. 1922. Botanical exploration in Bolivia. *Brooklyn Botanical Garden Record* 11:102ff.

Wolfes, O., and K. Rumpf. 1928. Über die Gewinnung von Harmin aus einer südamerikanischen Liane. *Archive für Pharmakologie* 266:188ff.

2

THE PSYCHOLOGY OF AYAHUASCA

CHARLES S. GROB, M.D.

The field of *ayahuasca* studies poses a challenge to mainstream psychiatry and psychology. Long neglected by Euro-American science, this Amazonian plant hallucinogen concoction known in native Quechua as the "vine of the dead," or "vine of the soul," has recently begun to attract increasing degrees of interest. Over the last several years investigations into the basic psychopharmacology and physiology of this powerful herbal medicine have been initiated. Questions are beginning to be posed examining the potential of ayahuasca to facilitate states of healing. Cross-cultural anthropological perspectives on the import of indigenous belief systems to ayahuasca's mechanism of action are being validated as essential to fully understanding its unique range of effects. Rational science will now need to confront the dilemma of how to comprehend and make sense of an experience that moves beyond the realm of rational, linear thought.

The fields of psychiatry and psychology have never had an appreciable comfort level with the mind states of aboriginal peoples. Native cultures have often been disparaged and the technologies designed to induce ritual trance states either pathologized or ignored. Years past, during a time of psychoanalytic preeminence, the medicine men, or healers, of these aboriginal peoples were judged to be mentally ill (Devereux

1958), their behaviors variably attributed to diagnoses ranging from schizophrenia to hysteria and epilepsy. The primitive medicine man, or shaman, was often identified as a deranged aboriginal tyrant and the wellspring of that psychopathology inflicting the entire tribal group, preventing their elevation into civilized society. Until quite recently the prevailing perception of the aboriginal has been one of the ignorant, deluded, and dangerous savage, whose only salvation lay in abandoning the traditions of his ancestors for the customs and beliefs of modern culture. The proposition of taking seriously the plant technologies underlying the collective belief systems found in native shamanism was given little credence by mainstream science and medicine.

Quietly, the tides of change have begun to turn. Heightened interest in the ancient tools of transcendence have surfaced. After decades of silence following the decline and virtual disappearance of hallucinogen research in the 1960s (Grob 1994), a period of revived interest within the fields of anthropology, psychiatry, and psychology has arisen. Additional activity, replaying some of the drama of that previous era of investigation, has occurred outside the bounds of formal and credential-bound science. Ethnobotanical explorations in diverse geographic regions have yielded a surprising plethora of psychoactive plants, some with no prior history of cultural identification. Knowledge of potent psychochemical recipes have begun to disseminate, often with the aid of the Internet. Use of plant hallucinogens, in both underground and formal settings, is growing. It is time for postmodern medical science to reawaken and be attentive to this rapidly emerging phenomenon. Beyond the need to assess safety parameters, the full implications to paradigms of healing and reality need to be grappled with.

We are on the threshold of a new era of scientific exploration. The ancient tools of consciousness alteration may yet hold valuable information needed to safely guide our culture into the next century and an uncertain future. For the mainstream fields of psychiatry and psychology, it is time to reopen the question of the value of shamanic ritual experience, and to explore the ceremonial structures and technologies within which these phenomena are embedded. From the cultural context of our own historical evolution, we must ask the question, What can we learn from the past and peoples with vastly different worldviews than our own? One valuable area of study that may provide answers is the actual psychoactive plant sacraments used in aboriginal rituals, their botanical identification, chemical constituents, pharmacologic effect, and cross-cultural function. Plant hallucinogens have a world-wide distribution and pan-historic record of application (Schultes and Hofmann 1992). There are approximately 150 different species of plant hallucinogens, 130 of these being primarily located in the New World, and only about 20 in Europe. Their role in early cultures is poorly understood, though some have speculated that they played a pivotal role at the inception of many of the worlds' major religions (Huxley 1977; Wasson et al. 1986). The areas of the world most richly endowed with these powerful botanicals are the tropical rain forests, particularly the Amazon Basin of northwest South America. And the prototype Amazonian plant hallucinogen concoction is ayahuasca.

WHAT IS AYAHAUSCA?

There is an important hallucinogenic drink consumed by the native peoples of the western half of the Amazon valley and by isolated tribes on the Pacific slopes of the Ecuadoran and Columbian Andes. The primary plant material used for this botanical brew is the bark of the giant forest liana *Banisteriopsis caapi*. This ayahuasca, as it is most commonly called, is prepared by boiling *Banisteriopsis* bark with the leaves of one or more admixture plants, most commonly *Psychotria viridis*. Although some aboriginal tribes have traditions of using *Banisteriopsis* solely, without any admixture plants, the predominant pattern of use has called for the addition of plants with diverse profiles of chemical constituents and psychoactive effects.

Ayahuasca possesses a unique phytochemistry. When taken alone, without the *Banisteriopsis*, many of the admixture plants are not psychoactive. Although *Psychotria viridis* is rich in alkaloids of the potent hallucinogen dimethyltryptamine (DMT), it is rendered biochemically inactive after oral consumption through inactivation by monoamine oxidase, an enzyme that degrades DMT, along with endogenous neurotransmitters. However, when *Psychotria* is prepared along with *Banisteriopsis*, which has monoamine oxidase inhibiting action, the

DMT is actively absorbed, passes the blood-brain barrier, and exerts powerful hallucinogenic effects on the central nervous system. How the native peoples of the Amazon discovered this sophisticated synergistic plant biochemistry is unknown, although the reductionistic explanation asserts that through generations, even centuries, of trial and error sampling of the abundant and diverse tropical flora, the aboriginal inhabitants of the region happened upon this unusual combination. Asking the native peoples themselves, however, yields a very different response. Virtually all of the ayahuasca-using tribes of the Amazon Basin, as well as the modern syncretic churches, who use this plant hallucinogen concoction as a legal psychoactive ritual sacrament, attribute the discovery of ayahuasca, along with the mythological origins of their own idiosyncratic religious belief systems, to a form of divine intervention. However human beings happened to come upon or were directed to this unique phytochemical combination, its discovery was integral to both the development of early native cultures as well as the rise of interest in these sacred plants in our own day.

Biochemically, ayahuasca is a combination of dimethyltryptamine, from the leaves of *Psychotria,* and three harmala alkaloids from *Banisteriopsis* bark, harmine, harmaline, and tetrahydroharmine. Dimethyltryptamine has a remarkable structural similarity to the endogenous neurotransmitter serotonin, or 5-hydroxytryptamine. The serotonin system is thought to be the primary neurotransmitter system involved in the activation and modulation of DMT's profound effect upon central nervous system function. The central structural feature of DMT biochemistry is the presence of the indole ring, a characteristic that it shares with serotonin as well as other potent hallucinogens, including lysergic acid diethylamide (LSD). Hallucinogenic tryptamines are present in a variety of plants, including orally active psilocybin mushrooms (the *teonanácatl* of the ancient Aztecs) and the virola snuffs of Amazonian aboriginals. Dimethyltryptamine, as well as its analogue 5-methoxy-dimethyltryptamine, rapidly induce an intense altered state of consciousness of relatively short duration when smoked or snorted intranasally but are entirely without effect when ingested orally. The harmala alkaloids derived from *Banisteriopsis* are ß-carbolines, which possess potent monoamine oxidase inhibiting (MAOI) action. This

MAOI activity allows for a unique profile of biochemical effects, most prominently the activation and augmentation of additive plants. When taken alone, the harmala alkaloids may induce varying degrees of hallucinogenic inebriation and are utilized ritually for this purpose by particular native tribes. The characteristically bitter taste of ayahuasca, along with its nauseating and emetic effects, are attributable to these harmala alkaloids.

HISTORICAL ANTECEDENTS

Archeological evidence supports the existence of the ritual use of plant hallucinogens by native peoples of the New World, long predating the arrival of the European explorers and colonists (Adovasio and Fry 1976; Torres et al. 1991). By the sixteenth century, however, particularly in the more tropical regions of Central and South America, with their greater abundance of psychoactive botanicals, the aboriginal utilization of native pharmacopeia began to evoke a harsh and punitive reaction. The occupying Spaniards and Portuguese, possessors now of most of the New World's rain forests, brutally persecuted and exploited native cultures (Taussig 1987). Observing the utilization of sacred plants to induce an ecstatic intoxication, and identifying the central role they played in aboriginal religion and ritual, these new European overlords harshly condemned their use. Hernando Ruiz de Alarcon, an early Spanish chronicler of native customs, described how the plants "when drunk deprive of the senses, because it is very powerful, and by this means they communicate with the devil, because he talks to them when they are deprived of judgement with the said drink, and deceive them with different hallucinations, and they attribute it to a god they say is inside the seed" (Guerra 1971). Condemned by the Holy Inquisition in 1616, the ceremonial use of plant hallucinogens by aboriginal peoples of the New World survived only by going deeply underground, remaining hidden from the hostile and rapacious European-imposed dominant culture.

Knowledge of ayahuasca use by native peoples of the Amazon was first recorded in the seventeenth century when Jesuit priests described the existence of "diabolical potions," prepared from forest vines by the native people of Peru (Ott 1994). There was no additional mention in the collective literature until the mid-nineteenth century when Richard Spruce, an English botanist doing fieldwork in the Amazon, identified the bark of a liana as the central substance in the legendary aboriginal *caapi*, or *yagé*. Later, when subjected to modern biochemical analysis, this specimen of Spruce's, having been kept for over one hundred years in storage at a British museum, was identified as a sample of *Banisteriopsis* containing harmala alkaloids. The first written report describing ayahuasca from a contemporaneous source was in 1858 by Manuel Villavicencio, an Ecuadoran geographer, who studied tribal groups in the Rio Napo area of Ecuador. Intrigued by the use of yagé in rituals, Villavicencio sought the opportunity to investigate firsthand the subjective effects of this beverage. He would later describe:

> in a few moments it begins to produce the most rare phenomena. Its action appears to excite the nervous system; all the senses liven up and all faculties awaken; they feel vertigo and spinning in the head, followed by a sensation of being lifted into the air and beginning an aerial journey; the possessed begins in the first moments to see the most delicious apparitions, in conformity with his ideas and knowledge; the savages [apparently the Záparo of eastern Ecuador] say that they see gorgeous lakes, forests covered with fruit, the prettiest birds who communicate to them the nicest and the most favorable things they want to hear, and other beautiful things relating to their savage life. When this instant passes they begin to see terrible horrors out to devour them, their first flight ceases and they descend to earth to combat the terrors who communicate to them all adversities and misfortunes awaiting them. . . . As for myself I can say for a fact that when I've taken ayahuasca I've experienced dizziness, then an aerial journey in which I recall perceiving the most gorgeous views, great cities, lofty towers, beautiful parks, and other extremely attractive objects; then I imagined myself to be alone in a forest and assaulted by a number of terrible beings from which I defended myself; thereafter I had the strong sensation of sleep. . . . (Villavicencio 1858; Harner 1973a).

ANTHROPOLOGICAL PERSPECTIVE

Among the diverse native peoples of the Amazon basin, ayahuasca has been known by a variety of names, including *caapi, yagé, natéma, mihi, kahi, pinde,* and *dapa.* Ayahuasca is perceived as a magic intoxicant, of divine origin, which facilitates release of the soul from its corporeal confinement, allowing it to wander free and return to the body at will, carrying with it information of vital import (Schultes and Hofmann 1992). Among native peoples, ayahuasca was traditionally used for purposes of magic and religious ritual, divination, sorcery, and the treatment of disease (Dobkin de Rios 1972). Living among the Cashinahua of the Peruvian Amazon, the anthropologist Kenneth Kensinger (1973) identified ayahuasca induced visions as integral to the genesis of volitional behavior. These visions are perceived as the experiences of an individual's "dream spirit," which has access to knowledge contained in the supernatural realms. Kensinger has described how the Cashinahua use ayahuasca as a means of receiving information not available through the normal channels of communication, which, along with additional information accessed through other means, constitute the basis for personal action. For the Cashinahua, however, taking ayahuasca is an unpleasant and even fearful experience, which is resorted to only when such revelations from the spirit world are urgently needed.

For the Jivaro Indians of the Ecuadoran Amazon, the supernatural realm, which can only be accessed through the door of the ayahuasca-induced experience, is seen as the true reality, whereas normal waking life is simply an illusion. Michael Harner, an anthropologist who has lived among and studied the Jivaro, has understood the importance of the ayahuasca experience itself in order to fully grasp the aboriginal mindset. Although traditional anthropological observers have typically assumed a passive role in compiling and recording native habits and customs, Harner elected to undergo a firsthand encounter with the object of his study. Harner has described how

> for several hours after drinking the brew, I found myself, although awake, in a world literally beyond my wildest dreams. I met bird-headed people, as well as dragon-like creatures who explained that they were the true gods of this world.

I enlisted the services of other spirit helpers in attempting to fly through the far reaches of the Galaxy. Transported into a trance where the supernatural seemed natural, I realized that anthropologists, including myself, had profoundly underestimated the importance of the drug in affecting native ideology (Harner 1973b).

Integral to the content of the realm of ayahuasca-induced visions is the cultural context in which they occur. Throughout the tropical rain forests of South America, the traditional ayahuasca-using tribal people have many shared common cultural elements as well as similar contextual themes for their mythologies and ayahuasca-induced experiences. Indeed, some anthropological observers have asserted that it is virtually impossible to separate the nature of the ayahuasca experience from its cultural context (Harner 1973a). Through contacting the supernatural realm of their ancestors, as well as their mythological deities and spirits, the ritual use of ayahuasca has served to bind the communities of disparate individuals into a cohesive collective culture. Culturally syntonic visions are induced by shamanic manipulation of set and setting to provide revelation, blessings, healing, and ontological security for those using such sacramental plants (Grob and Dobkin de Rios 1992). Within traditional contexts, shamanic initiation into the world of plant hallucinogens, including ayahuasca, have included long preparations associated with strict self-discipline, including prolonged social isolation, sexual abstinence, and diets free of meat, salt, sugar, and alcohol. The collective ingestion of these ceremonial sacraments by the adult members of the community achieves an amplified degree of social cohesion and identity that has been characterized by Mircea Eliade as a periodic symbolic regression to the "powerful time" of mythical origin (Eliade 1964). Ultimately, the intent of the collective ayahuasca sessions is not necessarily to evoke visions from each tribal individual's unconscious, but rather to assimilate and absorb the unconscious biographical personality structure into the cultural patterns of the visionary motifs (Andritzky 1989).

CONTEMPORARY CONTEXTS

Since the time of the European conquest and colonization of the New World several hundred years ago, the original native peoples of the Amazon basin have been all but eradicated. First through policies of virtual genocide, enslavement, and forced labor (Taussig 1987), then through the rampant dissemination of infectious disease, and later through the process of gradual acculturation to European values, culture, and religion, aboriginal customs and belief systems have barely survived to the present day. Only in the most remote regions of the rain forest have isolated tribes been able to avoid the relentless onslaught of modern culture, whether it has been from capitalist or missionary motive. Consequently, the traditional use of ayahuasca by native peoples has virtually disappeared, replaced by belief systems and forms of worship antithetical to the ways of their ancestors. By bringing such trappings of modern civilization to the world of the heathen, often in the form of missionary medical aid, the aboriginal people of the Amazon have been encouraged, sometimes gently and sometimes through coercion, to abandon their traditional use of ayahuasca. Nevertheless, knowledge of this extraordinary herbal concoction, representative of what Shultes and Hofmann (1992) have termed these "plants of the Gods," has survived. In diverse forms, reflecting their modern cultural contexts, the use of ayahuasca for purposes of healing and religious worship has persisted to our present day.

As the use of ayahuasca was central to conceptualizing and treating illness among native peoples, so this model for healing has persisted among mestizo, or mixed race, populations throughout the Amazon basin. The anthropologist Marlene Dobkin de Rios has studied the phenomena of urban mestizo healing in the Peurivan tropical rain forest city of Iquitos (Dobkin de Rios 1972). Her work has described not only a particular context in which the use of ayahuasca has persisted but also how it has served a protective function against the stressors of modern life. Demoralized by the shock of acculturative forces, mestizo inhabitants of Iquitos suffer from high rates of stress-induced anxiety and associated psychosomatic disorders. The use of ayahuasca, as supervised by mestizo healers, is incorporated into complex ritual healing ceremonies. Such a model for ayahuasca healing is most efficacious with illnesses that are believed to be magical in origin. The primary function of taking ayahuasca in this context is to diagnose the magical cause of illness, as well as to deflect and neutralize the evil spirits or magic that are believed to be responsible for particular types of illness. Within the context of a collective belief system that still maintains a conviction that such supernatural realms do exist, this model for ayahuasca healing among modern day Amazonian mestizos appears to function with a fair degree of success (Dobkin de Rios 1984).

Another model for ayahuasca healing has been developed in the Peruvian upper Amazon province of San Martin. The Takiwasi Center, created in the mid-1980s by a group of French and Peruvian physicians and healers, was inspired both by knowledge of the salutary effects of the legendary vine and the pressing need to identify an effective treatment for the rapidly increasing problem of coca paste addiction. A cheap intermediary between the raw coca plant and refined cocaine, coca paste is a highly addictive drug that has spread throughout the coca-cultivating and cocaine-producing regions of South America. Conventional treatment models for coca paste addiction have largely proved ineffectual, thus compounding this public health dilemma. Studying models for shamanic healing among extant native peoples, the Takiwasi group has developed their own involved structure for the ayahuasca-facilitated treatment of drug addiction (Mabit 1988; Mabit et al. 1995). The phenomena of Takiwasi may serve as an early example of the identification and use of ayahuasca for purposes of healing refractory conditions by physicians trained in modern medicine.

During the twentieth century the use of ayahuasca has been incorporated into the ritual practices of organized syncretic churches in Brazil. The first such ayahuasca church, the Santo Daime, began in the 1930s in the Brazilian Amazon region of Acre (Groisman and Sell 1995). Founded originally among mixed race *caboclos* and rubber tappers, the Santo Daime subsequently spread among middle-class populations throughout Brazil and ultimately abroad. The primary centers of Santo Daime activity included the founding of several self-sustaining communities in isolated regions of the Amazonian rain forest.

In the early 1960s, also in Acre state, an independent ayahuasca-using religion was founded by Jose Gabriel da Costa, who had acquired knowledge of this powerful sacramental plant preparation while working as a rubber tapper in the rain forests of Bolivia. Returning to his native Brazil, Mestre Gabriel, as he came to be called by his disciples, founded the Centro Espirita Beneficente União do Vegetal (UDV). Spreading primarily to urban areas, the UDV became the largest and most organized of the ayahuasca churches, ultimately establishing its headquarters in the Brazilian capital city of Brasilia (Ott 1993). The UDV was also primarily responsible for the successful petition to the Brazilian government to remove ayahuasca from the list of banned substances. Establishing an extraordinary precedent, the Brazilian government in 1987 declared ayahuasca to be a legal substance when used within the context of religious practice, thus becoming the first nation worldwide in almost 1600 years to allow the use of plant hallucinogens for spiritual purposes by its nonindigenous inhabitants.

Over the past decade knowledge and use of ayahuasca has spread throughout Europe and North America. This activity has come primarily from two directions. The Brazilian ayahuasca churches, the Santo Daime in particular, have established centers in many cities across Europe, with greatest activity in Spain, Holland, and Germany. The UDV has been much more circumspect and cautious, however, maintaining a relatively low profile and avoiding unnecessary and unwelcome media attention. This pattern has continued in the United States, particularly on the West Coast, where in recent years the Santo Daime has held a number of "works," generally open events with minimal screening or preparation of participants, whereas UDV activities, under direction from the centralized church hierarchy in Brazil, have limited participation only to formal UDV members and individuals who had been previously introduced to the plant sacrament and ritual.

Increasing interest in the ayahuasca studies of North American scholars and scientists has also begun to spur new activities in this area. In particular, the work of the autodidact writer and underground investigator Jonathan Ott has yielded a rich harvest of information. In books (Ott 1993; Ott 1994a) and articles (Ott 1994b; Ott 1999), Ott has, in voluminous detail, described the varieties of plant and synthetic substances with chemical properties similar to those of the primary plants of Amazonian ayahuasca, *Banisteriopsis* and *Psychotria*. These ayahuasca analogues, plants containing ß-carboline harmala alkaloids, including *Peganum harmala* (Syrian rue), along with additional plants rich in tryptamines (e.g., *Anadenanthera colubrina, Mimosa hostilis,* and *Philaris arundinacea*), provide rich, alternative nontropical sources of powerful plant hallucinogenic experience. With the virtually world-wide and previously largely unknown distribution of such ayahuasca analogues, Ott has provided the blueprint for what he has termed the pan-Gaean entheogenic revival (Ott 1994a). Similarly, laboratory ana-logues for ayahuasca, combinations of synthetic dimethyltryptamine, harmine, and harmaline, have been described that approximate the subjective effects of the prototype plant beverage. This pharmahuasca, as identified by Ott and others (Ott 1999; Callaway 1994), possesses the extraordinary capacity, along with the diverse plant analogues, to replicate the legendary ayahuasca experience for large numbers of "psychonauts" without proximity or access to the tropical plants. The implications of this phenomena, in an age of increasingly rapid dissemi-nation of information, make the medical and scientific examination of ayahuasca all the more compelling.

WHAT IS THE AYAHUASCA EXPERIENCE?

As is the case with all hallucinogens, the ayahuasca experience is pro-foundly affected by the extrapharmacological factors of set and setting (Bravo and Grob 1989). Intention, preparation, and structure of the session are all integral to the content and outcome of any encounter with hallucinogens, a clear distinction from virtually all other psy-chotropic agents. The diligent attention to these factors are known to be integral to the shamanic model of altered states of consciousness, minimizing risks and enhancing the likelihood of salutary results. The failure to adequately comprehend and adhere to the wisdom behind these time tested safeguards, on the other hand, often leads to the unfortunate consequences frequently observed within the context of contemporary recreational drug use and abuse (Grob and Dobkin de Rios 1992; Dobkin de Rios and Grob 1994).

Altered states of consciousness, including those induced by halluci-nogens, possess a variety of common elements (Ludwig 1969; Dobkin de Rios 1972). Before examining those features more closely identified with the ayahuasca experience, these shared properties merit review. The ten general characteristics understood to be virtually universal to such altered state experience include:

1. Alterations in Thinking. To varying degrees, subjective changes in concentration, attention, memory, and judgment may be induced in the acute state, along with a possible diminution or expansion of reflective awareness.

2. Altered Time Sense. The sense of time and chronology may become altered, inducing a subjective feeling of timelessness, or the experience of time either accelerating or decelerating. Time may be experienced as infinite, or infinitesimal in duration.

3. Fear of Loss of Control. An individual may experience a fear of losing his hold on reality or his sense of self-control. In reaction, increased resistance to the experience may occur, causing an amplification of underlying anxiety. If there is a positive cultural conditioning and understanding of the experience, mystical and positive transcendent states may ensue.

4. Changes in Emotional Expression. Along with reduction in volitional or conscious control, intense emotional reactivity may occur, ranging from ecstasy to despair.

5. Changes in Body Image. Alterations in body image are frequently reported, often associated with dissolution of boundaries between self and others and states of depersonalization and derealization where the usual sense of one's own reality is temporarily lost or changed. Such experiences may be regarded as strange and frightening, or as mystical, oceanic states of cosmic unity, particularly when sustained within the context of belief systems conditioned for spiritual emergent encounters.

6. Perceptual Alterations. Increased visual imagery, hyperacuteness of perceptions and overt hallucinations may occur. The content of these perceptual alterations are influenced by cultural expectations, group influences, and individual wish-fulfillment fantasies. They may reflect the psychodynamic expression of underlying fears or conflicts, or simple neurophysiologic mechanisms inducing geometric patterns

and alterations of light, colors, and shapes. Synesthesias, the transformation of one form of sensory experience into another, such as seeing auditory stimuli, may be experienced.

7. Changes in Meaning or Significance. While in a powerful altered state of consciousness, some individuals manifest a propensity to attach special meaning or significance to their subjective experiences, ideas, or perceptions. An experience of great insight or profound sense of meaning may occur, their significance ranging from genuine wisdom to self-imposed delusion.

8. Sense of the Ineffable. Because of the uniqueness of the subjective experience associated with these states and their divergence from ordinary states of consciousness, individuals often have great difficulty communicating the essence of their experience to those who have never had such an encounter.

9. Feelings of Rejuvenation. Many individuals emerging from a profoundly altered state of consciousness report a new sense of hope, rejuvenation, and rebirth. Such transformed states may be short-term or, conversely, may lead to sustained positive adjustments in mood and outlook.

10. Hypersuggestibility. While in the throes of altered state experience, individuals experience an enhanced susceptibility to accept or respond uncritically to specific statements. Nonspecific cues, reflecting cultural belief systems or group expectation, may similarly assume directives of weighty importance. The position of shaman, or session facilitator, particularly within the context of hallucinogen use, consequently becomes a role with great vested responsibility, as individual participants are highly susceptible to verbal and nonverbal input directed toward them. The content and outcome of such altered states experiences are often directly attributable to the integrity and skill of the leader.

Reports of specific ayahuasca effects vary greatly depending upon the cultural context, which may range from traditional native Amazonian ritual, to mestizo healing ceremony, to syncretic religious structure, to inquisitive Euro-American psychonautic exploration. The Tukano tribe of the Columbian Amazonia separate the ayahuasca experience into three stages. The first stage, which may begin within minutes of ingestion, induces the characteristic gastrointestinal reaction of nausea, emesis, and diarrhea, along with sweating, a sense of "flying," and the visual perception of vivid, kaleidoscopic array of brightly colored lights and geometric patterns. During the second phase of ayahuasca intoxication for the Tukano, the perception of brightly colored geometric patterns start to fade, while the sensation of flight into deep internal space intensifies along with an envisioning of three-dimensional forms of mythological and "monstrous" animals. The third and final stage involves the deepening of hallucinations, along with a progression into calmer and more peaceful visions and thought associations (Reichel-Dolmatoff 1975; Schultes and Winkelman 1995).

Commonalities of the ayahuasca visionary experience among diverse aboriginal groups of the tropical rain forests of South America have been described (Harner 1973b). Shared elements include:

1. The perception of the separation of the soul from the physical body associated with the sensation of flying. This astral voyage persists for the duration of the effects of the ayahuasca, after which the soul reenters the body. With some tribal people, the soul leaves the body in the form of a bird, which flies to some predetermined destination. A sensation of vertigo, or spinning, is often experienced.

2. Visions of snakes and jaguars, along with other predatory animals of the rain forest. Snakes and giant anacondas in particular populate the visions of native people. On some occasions, the ayahuasca voyager perceives himself to be attacked and consumed by these reptilian marauders, whereas at other times it is the snakes that are ingested. Confrontations with such rain forest predators may be empowering to the traveler, allowing him to forge an alliance with them as powerful shamanic spirit animals that may assist him in his journeys and battles in the supernatural realm.

3. The sense of contact with supernatural realms. Visions are experienced of deities and/or demons, consistent with the shamanic belief systems of native voyagers. For native Indians influenced by the entreaties of missionary Christianity, yet who have maintained

their traditions of ritual ayahuasca use, visitations to heaven and the realms of hell are reported.

4. Visions of distant persons, cities, and landscapes, which are understood by native Indians as clairvoyant experience. Such sensations of "seeing" events and locations far removed in place and time are utilized for purposes ranging from identifying where in the forest plentiful game may be available for the hunt, to inquiring about the health and well-being of relatives or friends, to practicing the many forms of sorcery that exist among aboriginal Amazonian tribes.

5. The sensation of "seeing" the detailed reenactment of recent unsolved crimes, or of identifying through visions the shaman responsible for bewitching a sick or dying person. Among many tribal groups, illness is understood to be caused by the actions of bewitchment, whereas the healing process relies upon the identification of the shaman responsible for the condition. Ayahuasca divination may address such tasks as identifying the perpetrators of recent homicides or thefts, discovering the plans of enemy attacks, predicting the imminent arrival of strangers, the adjudication of quarrels or disputes, and the investigation of whether spouses are faithful.

Modern observers have reported a range of experience reflective of their own cultural context, although rain forest motifs common to aboriginal peoples are often in evidence. Reports of predatory animals, particularly snakes, are described not only within the tropical setting of mestizo healing (Dobkin de Rios 1971; Flores and Lewis 1978), but by sophisticated inhabitants of urban areas far removed from the forest as well (Naranjo 1973). Depending upon the belief system of participants, both collective and individual, ayahuasca visionary experiences are shaped. Within the context of the Brazilian syncretic churches, for example, it is not uncommon for practitioners to encounter visions incorporating elements of their religious mythology, and to interpret these events according to church precept (Groisman and Sell 1995; Grob et al. 1996).

Although an element of unpredictability is inherent even with experienced users (Mabit et al. 1995), the likelihood of being overwhelmed by frightening visions is enhanced when individuals venture into the realm of ayahuasca with neither adequate preparation nor a ritual structure designed to contain and channel the experience. In 1960, anticipating the psychonautic exploration of many of his countrymen several decades later, the American poet Allen Ginsberg ventured into the Peruvian Amazon town of Pucallpa in search of the legendary yagé experience. Accessing a concoction of ayahuasca from a local *curandero,* or mestizo healer, Ginsberg ingested the brew. Describing his experience the next day in a letter written to his friend William Burroughs, Ginsberg wrote:

> First I began to realize my worry about the mosquitoes or vomiting was silly as there was the great stake of life and Death—I felt faced by Death, my skull in my beard on pallet and porch rolling back and forth and settling finally as if in reproduction of the last physical move I make before settling into real death—got nauseous, rushed out and began vomiting, all covered with snakes, like a Snake Seraph, colored serpents in aureole all around my body, I felt like a snake vomiting out the universe. . . . I was frightened and simply lay there with wave after wave of death—fear, fright, rolling over me till I could hardly stand it, didn't want to take refuge in rejecting it as illusion, for it was too real and too familiar—especially as if rehearsal of Last Minute Death my head rolling back and forth on the blanket and finally settling in last position of stillness and hopeless resignation to God knows what Fate—for my being—felt completely lost strayed soul—outside of contact with some Thing that seemed present—finally had a sense that I might face the Question there and then, and choose to die and understand—and leave my body to be found in the morning (Burroughs and Ginsberg 1963).

As is the case with other hallucinogens, ayahuasca has the innate potential to plunge those who might sample its range of experience into the depths of hell or, conversely, into the exalted planes of the

celestial realms. Contrasting with the terrifying existential nightmarish quality of Ginsberg's experience is the report of Heinz Kusel, a
trader living among the Chama Indians of northeast Peru during
the late 1940s. After two unpleasant episodes following ingestion of
ayahuasca, Kusel decided to undergo another attempt at achieving
that level of experience he had heard about from local inhabitants.
Describing the last of these three experiences taken under the super-
vision of Nolorbe, his native guide, Kusel would subsequently write
that:

> the images were not casual, accidental or imperfect, but fully
> organized to the last detail of highly complex, consistent, yet
> forever changing designs. . . . I was very conscious at the time
> of an inexplicable sensation of intimacy with the visions. They
> were mine and concerned only me. I remember an Indian tell-
> ing me that whenever he drank ayahuasca, he had such beauti-
> ful visions that he used to put his hands over his eyes for fear
> somebody might steal them. I felt the same way. . . . The color
> scheme became a harmony of dark browns and greens. Naked
> dancers appeared turning slowly in spiral movements. Spots of
> brassy lights played on their bodies which gave them the tex-
> ture of polished stones. Their faces were inclined and hidden in
> deep shadows. Their coming into existence in the center of the
> vision coincided with the rhythm of Nolorbe's song, and they
> advanced forward and to the sides, turning slowly. I longed to
> see their faces. At last the whole field of vision was taken up
> by a single dancer with inclined face covered by a raised arm.
> As my desire to see the face became unendurable, it appeared
> suddenly in full close-up with closed eyes. I know that when the
> extraordinary face opened them, I experienced a satisfaction of
> a kind I had never known. It was the visual solution of a per-
> sonal riddle (Kusel 1965).

NEUROPSYCHIATRIC RESEARCH
WITH AYAHUASCA

Scientific and medical interest in ayahuasca long predated recent activities in the field. During the 1920s and 1930s, European and American pharmacologists and physicians began to pay attention to the exotic plant drug concoction from the tropical rain forests of South America. Working with specimens of *Banisteriopsis* bark, believed at that time to be the sole constituent of the legendary jungle drink, the renowned German research psychopharmacologist, Louis Lewin, in his final project before his death, succeeded in isolating one of the active alkaloids of ayahuasca, harmine, which he initially named banisterine (Lewin 1929). At Lewin's suggestion, the neurologist Kurt Beringer explored the effects of banisterine, also known as telepathine, in reference to the legendary properties of the jungle vine, in the treatment of Parkinson's Disease. Administering the drug to fifteen patients with postencephalitic Parkinsonism, Beringer reported the dramatic improvement of the classic symptoms of rigidity and akinesia (Beringer 1928). Although the use of banisterine in the treatment of Parkinson's Disease would ultimately be replaced by other drugs, most notably L-dopa, this application provides an early reference point for the discovery of medicinal uses of ayahuasca by modern investigators (Sanchez-Ramos 1991).

Although notable gains have been made in the past two decades exploring the basic biochemistry of ayahuasca (McKenna et al. 1998), formal psychiatric research investigations have not been pursued until recently. In part attributed to the thirty-year-old taboo against pursuing sanctioned psychiatric research with hallucinogens (Grob 1994), and in part because of the long standing resistance to investigating primary plant products by modern science, the seemingly remote rain forest plant concoction ayahuasca has until now received virtually no attention by established medical scientists. The enthusiasm among European physicians earlier in this century following the breakthrough discoveries of Lewin and Beringer now long forgotten, medical-psychiatric ayahuasca investigation had long slumbered.

Even during the period of the 1950s and 1960s, when interest in the basic properties and even therapeutic potential of hallucinogens was considered a valid pursuit, little interest in conducting investigations of ayahuasca was in evidence. There was some activity, however, directed at exploring the effects of synthetic harmala alkaloids. Working with synthetic harmine, early investigators expressed doubt as to its inherent psychoactivity (Turner et al. 1955; Pennes and Hoch 1957). The first to establish that harmala alkaloids possessed hallucinogenic activity was the Chilean psychiatrist and psychopharmacologist, Claudio Naranjo. Examining the basic psychogenic properties of harmaline, Naranjo identified the hallucinogenic threshold to be at dosage levels above 1 milligram per kilogram (mg/kg) intravenously, or 4 mg/kg orally. Contrasting its effects to the classic hallucinogens LSD and mescaline, Naranjo described that "the typical reaction to harmaline is a closed-eye contemplation of vivid imagery . . . which is in contrast to the ecstatic heavens or dreadful hells of other hallucinogens" (Naranjo 1967). Naranjo found the effects of harmaline to be relatively calming and subtle, his subjects reporting relaxed states of philosophical and religious contemplation, without emotional turmoil. In eight of thirty patients to whom he administered the drug, pronounced ameliorations of neurotic symptoms were reported (Naranjo 1973). Although significant in establishing harmaline's basic psychoactivity, Naranjo was not able to account for the powerful affective and perceptual altering experiences described in reports of the effects of ayahuasca. Subsequent investigators would resolve this question by identifying that harmaline is not only present in relative trace amounts in ayahuasca—far exceeded quantitatively by harmine—but that ayahuasca is likely to possess its potent hallucinatory properties on the basis of admixture plants included in its preparation (McKenna et al. 1984; Ott 1993).

There exist in nature over seventy-five plants that have been used by indigenous peoples in the preparation of ayahuasca (Bianchi and Samorini 1993). Most commonly, admixture plants containing tryptamine derivatives, in particular N,N-dimethyltryptamine (DMT), are combined with ß-carboline harmala alkaloids (McKenna and Towers 1984) to induce the powerful hallucinogenic effect. During the 1950s and 1960s laboratory and clinical studies examined the psychopharmacologic effects of tryptamines, primarily as a probe to determine the neurobiologic basis of mental disease (Turner and Merlis 1959). Efforts to associate the presence of endogenous DMT with mental illness, however, did not prove successful owing to the detection of tryptamine derivatives in the body fluids of both patients and control subjects. Naturally occurring DMT has been detected in tissues and body fluids of several mammalian species, including humans (Callaway 1995). A close association between DMT and the serotonin (5-hydroxytryptamine) neurotransmitter system has also been identified. Indeed, endogenous production of DMT was noted to proceed directly from tryptamine. More recently, a novel hypothesis has been proposed that suggests that the natural occurrence and function of tryptamines within the human central nervous system can be explained by their role promoting the visual phenomenon of dream sleep (Callaway 1988). Similar to the mechanism of ayahuasca action, increased levels of endogenous ß-carbolines during sleep are presumed to facilitate activity of methylated tryptamines by blocking their metabolism.

During the 1990s, a program of clinical investigation of DMT in normal volunteer subjects was pursued at the University of New Mexico. The first hallucinogen research study to receive FDA approval after a greater than two decade hiatus, the team of R. J. Strassman and colleagues at the University of New Mexico studied basic pharmacologic, physiologic, and psychologic effects of DMT in normal volunteer subjects (Strassman et al. 1994). When administered graded doses of intravenous DMT in the clinical laboratory setting, subjects reported seeing with their eyes closed rapidly-moving, brightly-colored visual displays of images. Subjectively, they described experiencing a lessening of control initiated by a brief but overwhelming "rush," which led to a dissociated state where euphoria alternated or coexisted with anxiety. The dosage range for activation of the hallucinogenic state was determined to be between 0.2 mg/kg and 0.4 mg/kg intravenous DMT. At the dosage levels at or above 0.4 mg/kg subjects uniformly reported feeling overwhelmed by the intensity and speed of onset of the experience.

THE HOASCA PROJECT

The first, and to date only, formal psychiatric research investigation of ayahuasca took place in the Brazilian Amazon city of Manaus in 1993. A multinational collaborative study, the Hoasca Project, examined biochemical (Callaway et al. 1994), pharmacological (Callaway et al. 1996), physiological (Callaway et al. 1999), and psychiatric (Grob et al. 1996) perspectives. The primary objective of these investigations was to establish a core of qualitative and quantitative data on the psychopharmacology of ayahuasca, which could establish relative safety profiles for human consumption as well as provide the foundation for future studies (McKenna et al. 1998). Given the legally protected status of ayahuasca in Brazil, a unique environment for the sanctioned investigation of hoasca (the Portuguese transliteration of ayahuasca) was available. Conducted with the full cooperation of the syncretic religious church, União do Vegetal (UDV), the Hoasca Project has established a model and precedent for the biomedical investigation of ayahuasca in its natural setting.

Method

For the purposes of this pilot investigation, fifteen long-term users of ayahuasca and fifteen matched controls were recruited. Ayahuasca-using subjects were all members of the UDV, of at least ten years standing, who consumed ayahausca in religious rituals at a frequency of at least two times monthly. Control subjects were matched along all demographic parameters, with the exception that they did not belong to the UDV and had never consumed ayahuasca. A variety of parameters were utilized to assess past and current levels of psychological function. Both ayahuasca-experienced subjects and normal controls were administered structured psychiatric diagnostic interviews (Composite International Diagnostic Interview [CIDI]), life-story interviews, personality testing (Tridimensional Personality Questionnaire [TPQ]), and neuropsychological testing (WHO-UCLA Auditory Verbal Learning Test). Ayahuasca-experienced subjects were asked to fill out an additional questionnaire (Hallucinogen Rating Scale [HRS]) following an experimental ayahuasca session. Each of the ayahuasca subjects were also interviewed utilizing a semistructured format designed to ascertain their life stories. In addition to this psychiatric investigation, a research methodology designed to evaluate serotonin biochemistry, through the examination of platelet serotonin receptor activity in both experimental and control subjects, was pursued. Additional biological investigations included tryptamine and harmala pharmacokinetics as well as acute physiological and neuroendocrine effects of ayahuasca in long-term users.

Results

Diagnostic and life-story interviews identified appreciable past psychiatric and substance abuse histories in the UDV subjects prior to their entry into the ayahuasca church, including 73 percent with histories of significant alcohol use, 33 percent with alcohol binging associated with violent behavior, 27 percent with stimulant abuse, and 53 percent with tobacco dependence. For all of these subjects, however, past psychopathology had resolved following initiation and regular attendance at ayahuasca ceremonies. Personality testing identified significant differences between the ayahuasca-using and nonusing groups. These included measures of novelty seeking, with UDV members described as being more reflective, rigid, loyal, stoic, slow-tempered, frugal, orderly, and persistent, and also scoring higher on measures of social desirability and emotional maturity than controls. Ayahuasca-using subjects were also distinguished from controls on the harm avoidance domain as being more confident, relaxed, optimistic, carefree, uninhibited, outgoing, and energetic. Overall, the UDV ayahuasca using group scored higher on traits of hyperthymia, cheerfulness, stubbornness, and overconfidence than their nonusing counterparts. Baseline neuropsychological testing also revealed differences between the two groups, with the long-term ayahuasca users demonstrating significantly higher scores on measures of concentration and short-term memory. The final psychological instrument, employed on ayahuasca subjects only, was the Hallucinogen Rating Scale, designed to correlate the intensity and phenomenology of the subjective state with known measures of intravenous dimethyltryptamine (Strassman et al. 1994). In this study of ayahuasca, scores in the 0.1 to 0.2 mg/kg range of intravenous DMT were recorded.

Life-story interviews were employed to gather additional personal histories of UDV subjects prior to ayahuasca initiation, the nature of their first ayahuasca experience, and an account of how their lives had changed following entry into the syncretic ayahuasca church. For most of the interview sample, their lives before entry into the UDV were described as impulsive, disrespectful, angry, aggressive, oppositional, rebellious, irresponsible, alienated, and unsuccessful. Many of them had profound initial encounters with ayahuasca. A common theme for their visionary experience was the perception of being on a self-destructive path that would ultimately lead to an ignominious end unless they radically reformed their personal conduct and orientation. Descriptions of these frightening visions included "I had a vision of myself in a car going to a party. There was a terrible accident and I could see myself die." "I was at a carnival, on a carousel, going around and around without ever stopping. I didn't know how to get off. I was very frightened." "I could see where I was going with the life I was leading. I could see myself ending up in a hospital, in a prison, in a cemetery." "I saw myself arrested and taken to prison. They were going to execute me for a horrible crime I had committed." While in the throes of their nightmarish visionary experience, several of the subjects reported encountering the founder of the UDV, Mestre Gabriel, who would deliver them from their terrors: "I saw these horrible, ugly animals. They attacked me. My body was disassembled, different parts were lying all over the ground. Then I saw the Mestre. He told me what I would need to do to put all my body parts back together." "I ran through the forest terrified that I was going to die. Then I saw the Mestre. He looked at me. I was bathed in his light. I knew I would be okay." "I was in a canoe, out of control, going down the river. I thought I would die. But then I saw Mestre Gabriel in a canoe in front of me. I knew that as long as I stayed with the Mestre I was safe."

All of the long-term ayahuasca-using subjects reported during the life-story interviews that they had undergone a personal transformation following entry into the UDV and regular participation in ritual ayahuasca use. In addition to entirely discontinuing cigarette, alcohol, and recreational drug use, they reported a radical restructuring of their personal conduct and value systems. One subject described how: "I used to not care about anybody, but now I know about responsibility. Every day I work on being a good father, a good husband, a good friend, a good worker. I try to do what I can to help others. . . . I have learned to be calmer, more self-confident, more accepting of others. . . . I have gone through a transformation." Subjects emphasized the importance of "practicing good deeds," watching one's words, and having respect for nature. Subjects also reported sustained improvement in memory and concentration, persistent positive mood states, fulfillment in day-to-day interactions, and a sense of purpose, meaning, and coherence to their lives.

All of the subjects interviewed unequivocally attributed the positive changes in their lives to their experiences within the UDV and their participation in the ritual ingestion of ayahuasca. They described ayahuasca as a catalyst for their moral and psychological evolution. They also insisted, though, that it was not necessarily the ayahuasca alone that was responsible, but rather partaking of the ayahuasca within the ritual context of the UDV ceremonial structure. Criticism of other Brazilian groups that were said to use ayahuasca in less focused and less controlled settings, was expressed by some of the subjects. The UDV was portrayed as a "vessel" that enabled them to safely navigate the often turbulent states of consciousness induced by ayahuasca. They described the UDV as their "mother . . . family . . . house of friends," providing them with "guidance and orientation" and allowing them to walk the "straight path." They emphasized the importance of "união," or union, of the plants and of the people. Without the structure of the UDV, these subjects asserted, ayahuasca experiences may be unpredictable and lead to an inflated sense of self. Within the "house of the UDV," however, the ayahuasca-induced state is controlled and directed "down the path of simplicity and humility."

Discussion

The Hoasca Project has been the first formal psychobiologic investigation of the legendary Amazonian sacrament. Psychological profiles of long-term members of the Brazilian ayahuasca church, União do Vegetal, reveal high levels of function compared to normal controls, including healthier personality measures, superior neuropsychological function. Whether these findings are attributable to the direct effects of ritual ayahuasca ingestion or whether they were self-selecting factors for affiliating with such a process to begin with cannot be determined, given the methodological limitations of this pilot investigation, nevertheless a strong preliminary case can be made for ayahuasca's salutary effects when utilized in such a context. A key factor when examining the apparent outcome of frequent hallucinogen, or ayahausca, use is the set and setting, and the role of suggestibility. For the subjects studied, the syncretic church they belonged to provided a protective and supportive community, as is often the case with all forms of religion. What might mark the phenomenon of the ayahuasca church as distinct, however, is the utilization of a powerful plant hallucinogen as a psychoactive religious sacrament. Highly susceptible to both explicit messages and implicit collective belief systems, the individual under the influence of ayahuasca is in a psychological state of heightened suggestibility. Whether the effects of the experience will ultimately be salutary or not appear in large part to be determined by the content of the explicit and implicit messages conveyed and the integrity of the religious structure.

CONCLUSIONS

It is perhaps ironic that as we prepare to transition to a new century, and a new millennium, interest in the ancient arts of transcendence has begun to increase. From first contact, some five hundred years ago, the Europeans who came to the Americas scorned and demonized the psychoactive plants the indigenous peoples used in their healing practices and religious rituals. Not deemed worthy of serious investigation, plant hallucinogens such as ayahuasca remained of interest only to a handful of maverick anthropologists and ethnobotanists. Recently, however, efforts at initiating formal multidisciplinary study of ayahuasca have kindled hopes that rigorous evaluation of this Amazonian plant concoction may yield valuable new information about cross-cultural belief systems, the range of mental function, and novel paradigms for healing.

From a variety of perspectives, the study of ayahuasca represents an opportunity to advance our knowledge of the human condition and the myriad conditions that influence it. Through anthropological examination of the role such plant hallucinogens have played in indigenous cultures, we may determine how particular structures and cultural contexts channel experience and optimize safety. And through the utilization of state of the art research methodologies and advanced neuroscience technologies, critical new information may become available to the fields of medicine and psychiatry. From a biological reductionistic perspective, the effects of ayahuasca on the serotonergic system in particular may lead to a new understanding of neural mechanisms responsible for central nervous system function (Callaway et al. 1994; Grob et al. 1996; McKenna et al. 1998). New models for treatment utilizing this powerful plant hallucinogen will need to incorporate knowledge gleaned both from anthropological studies of set and setting as observed within shamanic practice, as well as psychobiologic investigations of neurotransmitter and neuroendocrine effects of ayahuasca to achieve optimal response.

As knowledge of the highly unusual and intriguing ayahuasca experience spreads, care will have to be taken to prevent its exploitation and unsafe utilization. Already, concern has been expressed about "drug tourism" in the Amazon involving North American and European travelers in search of adventure and novel experience (Dobkin de Rios 1994). Not only do individuals often place themselves in danger through casual experimentation with unknown brews in what can be an uncontrolled and unpredictable setting but also may have a detrimental influence upon the local cultures. And given the hypersuggestible effects of all hallucinogens, risks may also exist when individuals allow themselves to be guided by others with questionable integrity and limited expertise. Care will also have to be taken to establish pharmacologic safety parameters for ayahuasca use, with particular attention given to potential adverse interactions with conventional pharmaceutical medications (Callaway and Grob 1999).

The study of ayahuasca represents a challenge to mainstream culture through the phenomenon of new and novel forms of religious practice, exemplified by the ayahuasca churches of Brazil that have lately spread to North America and Europe. As with the case of other plant hallucinogens employed as religious sacraments, in particular the use of peyote by the Native American Church, vital questions regarding freedom of religious practice will have to be addressed. The health professions of modernity will also need to confront the challenges of responding to the potential salutary effects of this ancient plant medicine traditionally utilized within the context of healing paradigms long regarded as alien and inconsequential. Tentative evidence of ayahuasca's capacity to facilitate healing and recuperative responses in individuals inflicted with psychiatric disorders (Grob et al. 1996) and medical illness (Topping 1998) necessitate further rigorous research investigation to establish the validity of these preliminary findings. Questions of ayahuasca's putative role in the treatment of addictive disorders, antisocial behavior or even neoplastic disease are compelling questions awaiting further exploration. To accurately determine ayahuasca's potential value as a medicine, it will ultimately be necessary to move beyond the boundaries of conventional treatment models and incorporate the lessons learned by past and distant cultures. Only then, as ancient technologies of transcendence are embedded within modern research methodologies, will we discover what true value this mysterious vine of the soul may have to us and our descendants.

References

Adovasio, J. M., and C. F. Fry. 1976. Prehistoric psychotropic drug use in northeastern Mexico and trans-Pecos Texas. *Economic Botany* 20:94–96.

Andritzky, W. 1989. Sociopsychotherapeutic functions of ayahuasca healing in Amazonia. *Journal of Psychoactive Drugs* 21:77–89.

Beringer, K. 1928. Uber ein neues, auf das extrapyramidal-motorische System wirkendes Alkaloid (Banisterin). *Nervenarzt* 1:265–75.

Bianchi, A., and G. Samorini. 1993. Plants in association with ayahuasca. *Jahrbuch für Ethnomedizin* (Yearbook of Ethnomedicine) 2:21–42.

Bravo, G., and C. Grob. 1989. Shamans, sacraments and psychiatrists. *Journal of Psychoactive Drugs* 21:123–28.

Burroughs, W. S., and A. Ginsberg. 1963. *The Yagé Letters*. San Francisco: City Lights Books.

Callaway, J. C. 1988. A proposed mechanism for the visions of dream sleep. *Medical Hypotheses* 26:119–24.

———. 1994. Some chemistry and pharmacology of ayahuasca. *Jahrbuch für Ethnomedizin* (Yearbook of Ethnomedicine) 3:295–98.

———. 1995. DMTs in the human brain. *Jahrbuch für Ethnomedizin* (Yearbook of Ethnomedicine) 4:45–54.

Callaway, J. C., L. P. Raymon, W. L. Hearn, D. J. McKenna, C. S. Grob, G. S. Brito, and D. C. Mash. 1996. Quantitation of *N,N*-dimethyltryptamine and harmala alkaloids in human plasma after oral dosing with ayahuasca. *Journal of Analytical Toxicology* 20:492–97.

Callaway, J. C., D. H. McKenna, C. S. Grob, G. S. Brito, L. P. Raymon, R. E. Poland, E. N. Andrade, E. O. Andrade, and D. C. Mash. 1999. Pharmacology of hoasca alkaloids in healthy humans. *Journal of Ethnomedicine*, 65:243–56.

Callaway, J. C., and C. S. Grob. 1998. Ayahuasca preparations and serotonin re-uptake inhibitors: A potential combination for severe adverse interactions. *Journal of Psychoactive Drugs*, 30:367–69.

Devereux, G. 1958. Cultural thought models in primitive and modern psychiatric theories. *Psychiatry* 21:359–74.

Dobkin de Rios, M. 1971. Ayahuasca, the healing vine. *International Journal of Social Psychiatry* 17:256–69.

———. 1972. *Visionary Vine: Hallucinogenic Healing in the Peruvian Amazon.* San Francisco: Chandler Publishing.

———. 1984. *Hallucinogens: Cross-Cultural Perspectives.* Albuquerque: University of New Mexico Press.

———. 1994. Drug tourism in the Amazon. *Jahrbuch für Ethnomedizin* (Yearbook of Ethnomedicine) 3:307–14.

Dobkin de Rios, M., and C. S. Grob. 1994. Hallucinogens, suggestibility and adolescence in cross-cultural perspective. *Jahrbuch für Ethnomedizin* (Yearbook of Ethnomedicine) 3:123–32.

Eliade, M. 1964. *Shamanism: Archaic Techniques of Ecstasy.* Princeton, NJ: Princeton University Press.

Flores, F. A., and W. H. Lewis. 1978. Drinking the South American hallucinogen ayahuasca. *Economic Botany* 32:154–56.

Grob, C. S. 1994. Psychiatric research with hallucinogens: What have we learned? *Jahrbuch für Ethnomedizin* (Yearbook of Ethnomedicine) 3:91–112.

Grob, C. S., and M. Dobkin de Rios. 1992. Adolescent drug use in cross-cultural perspective. *Journal of Drug Issues* 22:121–38.

Grob, C. S., D. J. McKenna, J. C. Callaway, G. S. Brito, E. S. Neves, G. Oberlaender, O. L. Saide, E. Labigalini, C. Tacla, C. T. Miranda, R. J. Strassman, and K. B. Boone. 1996. Human psychopharmacology of hoasca, a plant hallucinogen used in ritual context in Brazil. *Journal of Nervous and Mental Disease* 184:86–94.

Groisman, A., and A. B. Sell. 1995. "Healing power": Cultural-neuro-phenomenological therapy of Santo Daime. *Yearbook Cross-Cultural Medicine* 6:241–55.

Guerra, F. 1971. *The Pre-Columbian Mind.* London: Seminar Press.

Harner, M., ed. 1973a. *Hallucinogens and Shamanism*. London: Oxford University Press.

———. 1973b. Common themes in South American indian yagé experiences. In *Hallucinogens and Shamanism*, ed. M. Harner, 155–75. London: Oxford University Press.

Huxley, A. 1977. *Moksha: Writings on Psychedelics and the Visionary Experience*. Los Angeles: J. P. Tarcher.

Kensinger, K. B. 1973. *Banisteriopsis* usage among the Peruvian Cashinuaha. In *Hallucinogens and Shamanism*, ed. M. Harner, 1–14. London: Oxford University Press.

Kusel, H. 1965. Ayahuasca drinkers among the Chama Indians of northeast Peru. *Psychedelic Review* 6:58–66.

Lewin, L. 1929. *Banisteria Caapi, ein neues Rauschgift und Heilmittel*. Berlin: Verlag von Georg Stilke.

Ludwig, A. M. 1969. Altered states of consciousness. In *Altered States of Consciousness*, ed. C. T. Tart, 11–24. New York: John Wiley and Sons.

Mabit, J. 1988. Ayahuasca Hallucinations Among Healers in the Peruvian Upper Amazon. Document de Travail. Lima: Instituto Frances de Estudios Andinos.

Mabit, J., R. Giove, and J. Vega. 1995. Takiwasi: The use of Amazonian shamanism to rehabilitate drug addicts. *Yearbook Cross-Cultural Medicine* 6:257–85.

McKenna, D. J., G. H. N. Towers, and F. S. Abbott. 1984. Monoamine oxidase inhibitors in South American hallucinogenic plants: Tryptamine and beta-carboline constituents of ayahuasca. *Journal of Ethnopharmacology* 10:195–223.

McKenna, D. J., J. C. Callaway, and C. S. Grob. 1998. The scientific investigation of ayahuasca: A review of past and current research. *Heffter Review of Psychedelic Research* 1:65–77.

Naranjo, C. 1967. Psychotropic properties of the harmala alkaloids. In *Ethnopharmacologic Search for Psychoactive Drugs*, eds. D. H. Efron, R. Holmstedt, and N. S. Klein, 385–91. U.S. Public Health Service Publication No. 1645. Washington, D.C.: GPO.

———. 1973. *The Healing Journey: New Approaches to Consciousness*. New York: Random House.

Ott, J. 1993. *Pharmacotheon: Entheogenic Drugs: Their Plant Sources and History*. Kennewick, WA: Natural Products.

———. 1994a. *Ayahuasca Analogues: Pangæan Entheogens.* Kennewick, WA: Natural Products.

———. 1994b. Ayahuasca and ayahuasca analogues: Pangæan entheogens for the new millennium. *Jahrbuch für Ethnomedizin* (Yearbook of Ethnomedicine) 3:285–93.

———. 1999. Pharmahuasca: Human pharmacology of oral DMT plus harmine. *Journal of Psychoactive Drugs* 31 (2): 171–77.

Pennes, H. H., and P. H. Hoch. 1957. Psychotomimetics, clinical and theoretical considerations: Harmine, win-2299 and nalline. *American Journal of Psychiatry* 113:887–92.

Reichel-Dolmatoff, G. 1975. *The Shaman and the Jaguar.* Philadelphia: Temple University Press.

Sanchez-Ramos, J. R. 1991. Banisterine and Parkinson's Disease. *Clinical Neuropharmacology* 14:391–402.

Schultes, R. E., and A. Hofmann. 1992. *Plants of the Gods: Their Sacred, Healing and Hallucinogenic Powers.* Rochester, VT: Healing Arts Press.

Schultes, R. E., and M. Winkelman. The principal American hallucinogenic plants and their bioactive and therapeutic properties. *Yearbook Cross-Cultural Medicine Psychotherapy* 6:205–39.

Strassman, R. J., C. R. Qualls, E. H. Uhlenhuth, and R. Kellner. 1994. Dose-reponse study of N,N-dimethyltryptamine in humans. *Archives of General Psychiatry* 51:98–108.

Taussig, M. 1987. *Shamanism, Colonialism and the Wild Man: A Study in Terror and Healing.* Chicago: University of Chicago Press.

Topping, D. M. 1998. Ayahuasca and cancer: One man's experience. *MAPS* 8 (3): 22–26.

Torres, C. M., D. B. Repke, and K. Chan. 1991. Snuff powders from prehispanic San Pedro de Atacama: Chemical contextural analysis. *Current Anthropology* 32:640–49.

Turner, W. J. et al. 1955. Concerning theories of indoles in schizophrenigenesis. *American Journal of Psychiatry* 112:466–67.

Villavicencio, M. 1858. *Geografía de la República del Ecuador.* New York: R. Craigshead.

Wasson, R. G., S. Kramrisch, J. Ott, and C. A. P. Ruck. 1986. *Persephone's Quest: Entheogens and the Origins of Religion.* New Haven, CT: Yale University Press.

3

Phytochemistry
and Neuropharmacology
of Ayahuasca

J. C. Callaway, Ph.D.

INTRODUCTION

Homo Sapiens, that determined biped, has embarked on an extensive course of trial and error to identify the medicinal plants, and the resulting information has been applied to our cultures just as any other important tool. It may be no accident, then, that our neurological receptors happen to accommodate such a wide variety of psychoactive alkaloids. Perhaps it is due to our collective exposure to these substances over time. These discoveries were probably made by accident, initially, then deliberately. The earliest experiences with these substances were probably repeated because of some adaptive significance, such as heightened visual acuity (McKenna 1992), for example. It is even possible that the continued use of psychoactive drugs had, in some way, led to the rapid development of the human neocortex as a response to this xenobiotic challenge (note the structural similarity between serotonin and psilocin in Figure 1, page 97, for example).

The study of human drug use, either for medical, social, recreational, or religious purposes, is *ethnopharmacology*. In this field of study the primary goal is to identify plants that have been useful in the development and maintenance of human cultures, and then identify their active components by phytochemical and pharmacological analyses. Another important goal of ethnopharmacology is to apply this knowledge to modern cultures. The modification of consciousness, in particular, is apparently an essential experience for the human psyche. Although such effects can be triggered by a wide variety of methods, the primary advantage of pharmacological means is in reliability.

PHYTOCHEMISTRY

Certain psychoactive plants have been called "plant teachers," because of the information they apparently impart to the user (Luna and Amaringo 1991). Most of these plants are now known to contain specific chemical entities that, if administered in their pure form at sufficient dosages, bring about analogous modifications in perception and awareness in most human beings. As these effects are increasingly identified with certain chemicals, it is just as appropriate to speak of them as "molecular teachers," for these molecules are the most fundamental forms of matter that can precipitate these effects. From the molecular level, psychoactive drugs are known to bind to neuronal receptors within the brain (Callaway and McKenna 1998). This action is correlated with profound changes in cognition, but little more can be said about the content of this highly personal, yet nearly universal, experience.

The plant teachers have been held in high regard since the dawning of time, and their sacred status has remained intact to this day precisely for these effects. Just over one hundred years ago, Arthur Heffter isolated four pure chemicals from a small cactus and systematically ingested each one, eventually ascribing the awareness expanding properties of peyote to a chemical that is now known as mescaline (Heffter 1896, 1898).

Aside from the active components of a natural product, countless minor components may serve to modify the central effects of a substance in unique ways. This is one reason why natural products tend to differ in effect from synthetic chemicals. Nevertheless, specific molecules are essentially responsible for the psychopharmacologic effects that were once known only from plants. Mescaline from *Lophophora*, ibogaine from *Iboga,* and THC from *Cannabis* are only a few of the many examples that can be listed. The psilocin (Figure 1) and psilocybin that are found in various species of fungi serve as a classical example of this point. After isolating and identifying these active components from *Psilocybe mexicana,* Albert Hofmann then synthesized them by organic chemical methods. The synthetic psilocybin, in the form of pills, was then brought to the remote village of Huatla de Jimenez in Oaxaca, Mexico, where Gordon Wasson first learned of the magic mushrooms from the Mazatec shaman María Sabina. After taking the pills, she confirmed that the substance was "the same God" she had previously known only through the sacred mushroom (Hofmann 1978).

NEUROPHARMACOLOGY

Part of the reason why psychoactive drugs effect normal psychoactivity is that they structurally resemble, on a molecular level, the neuroactive agents that our own bodies produce to fill the receptor sites that are found on nerve cells. These endogenous agents are called neurotransmitters, and they are responsible for communication between nerves by relaying electrical impulses from one cell to another. Dopamine, norepinephrine, and serotonin are the primary neurotransmitters that are thought to be involved with behavior. By analogy, many of the psychotropic drugs that are used for religious purposes tend to have an indole or tryptamine structure, like serotonin, or a phenethylamine structure, like dopamine and norepinephrine. It may be worth noting here that epinephrine, or adrenaline, which is a metabolite of norepinephrine and released by the adrenal glands, is fairly similar to methamphetamine in structure and activity (Figure 1).

It is truly a wonder why so many different forms of life produce and conserve these unique patterns of atoms like serotonin, which is found in all animals and some plants. The mystery is only deepened by the fact that substantially similar molecules do impart such profound effects on the human mind. Additional molecular structures of a few well known substances are presented in Figure 2 to allow more readers some perspective in this discussion. It is also important to keep in mind that organic chemistry is only the most recent and consistent system that has been devised to categorize and contemplate the unique entities.

Neurotransmitters:

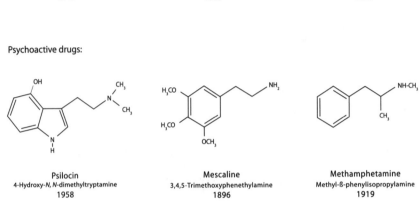

Serotonin	Dopamine	Epinephrine
5-Hydroxytryptamine	3,4-Dihydroxyphenethylamine	"Adrenaline"
1940	1923	1901

Psychoactive drugs:

Psilocin	Mescaline	Methamphetamine
4-Hydroxy-N, N-dimethyltryptamine	3,4,5-Trimethoxyphenethylamine	Methyl-ß-phenylisopropylamine
1958	1896	1919

Figure 1. Molecular structures of the neurotransmitters serotonin, dopamine, and epinephrine, compared with the structures of psilocin, mescaline, and methamphetamine. Also provided are the dates when these chemicals began to surface in the scientific literature, as a consequence of their unique properties on the human central nervous system.

SEROTONIN AS A NEUROTRANSMITTER AND PSYCHOACTIVE AGENT

The neurotransmitter serotonin, chemically known as 5-hydroxytryptamine, or 5-HT, is produced throughout the brain and gastrointestinal tract and seems to be responsible for higher functions of behavior, such as planning and other time-related events. It is produced in the body from *L*-tryptophan, an essential dietary amino acid (Figure 3, page 99).

In addition to serving as a neurotransmitter, serotonin is also the metabolic precursor of melatonin, which the human body makes during the night and when eyes are closed, as in some forms of meditation. Deficiencies in tryptophan and serotonin have been linked to mental

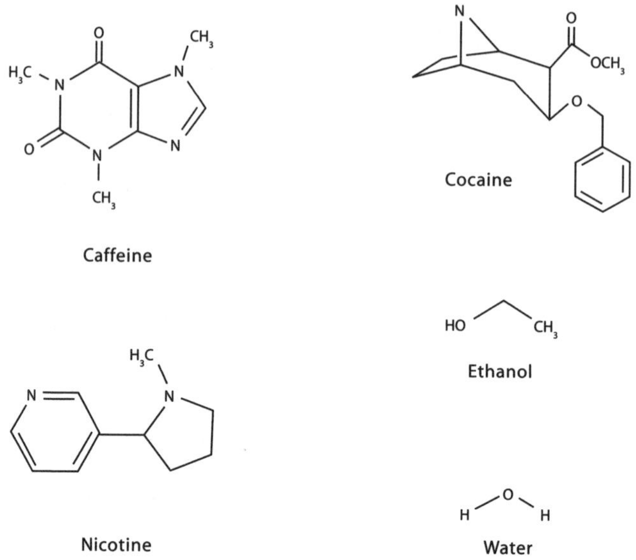

Figure 2. Molecular structures of other well known chemicals, illustrated for perspective and comparison.

disorders such as violent alcoholism, anxiety, depression, and suicide. Decreased serotonergic activity can stem from too little tryptophan entering the brain, too little serotonin produced from available tryptophan, aggressive uptake of serotonin from the synaptic cleft, or excessive activity of the enzymes that metabolize serotonin. Increased serotonergic activity, however, can be induced by pharmacologic means, and this is the primary goal of modern antidepressant medications.

Serotonin's major function is basically one of inhibition within the complex neurochemical pathways of the central nervous system (CNS), as if to screen out spurious bits of data to allow one to better focus on the task at hand. Modifying the action of serotonergic functions within a living organism typically results in observable changes in behavior. Many psychotropic substances, whether purified synthetic powders or crude natural products, affect at least some aspect of serotonergic activity.

Most serotonin is eventually metabolized to 5-hydroxyindoleacetic acid (5-HIAA) by monoamine oxidase (MAO), an enzyme in the body that also deactivates other neurotransmitters by a specific oxidation reaction. While the other neurotransmitters may use additional metabolic pathways, serotonin is fairly dependent on the activity of MAO

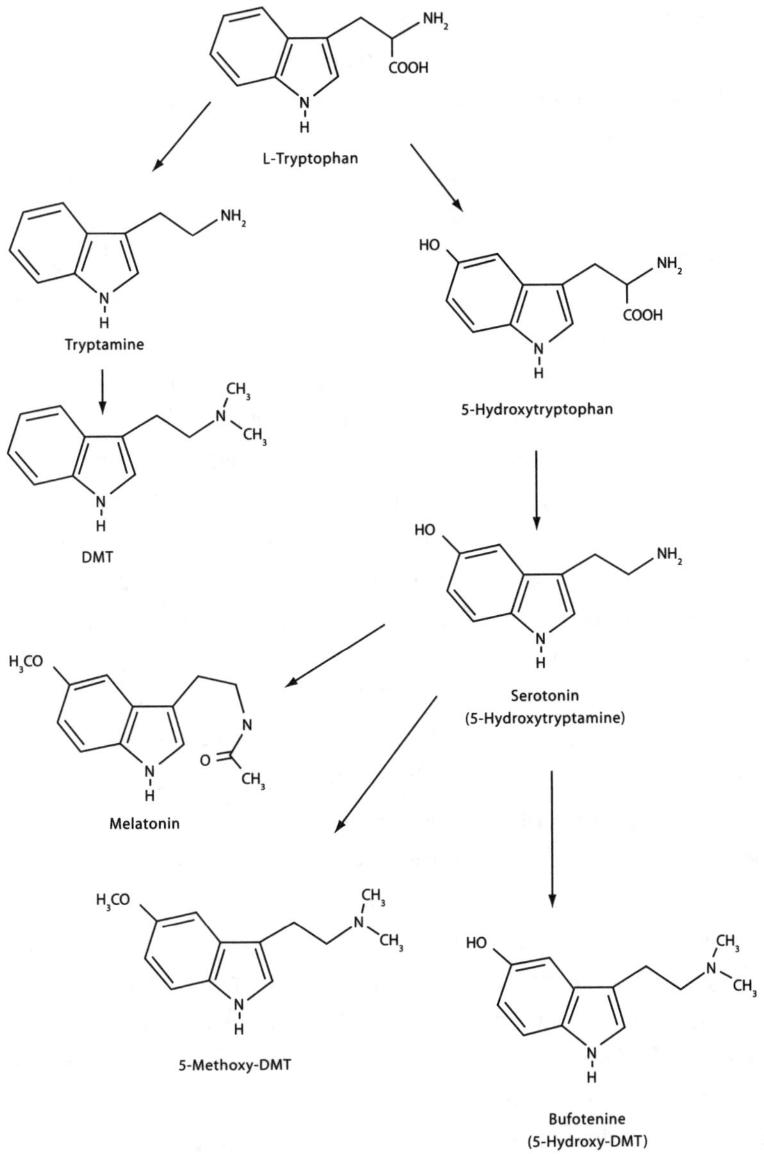

Figure 3. Metabolic pathways for the production of several endogenous indoles from dietary tryptophan.

for its deactivation. Thus, when MAO is inhibited, levels of serotonin in the brain begin to increase as its production continues unabated. This action alone can have noticeable psychoactive effects as the brain becomes hyperactivated by its own neurotransmitter.

Modern antidepressant drugs exploit this same mechanism (i.e., MAO inhibition) to increase neurotransmitter levels in depressed individuals. Newer antidepressant medications attempt to achieve the same goal by selectively blocking the recycling mechanism of serotonin (i.e., its reuptake), which encourages more of this neurotransmitter to seek available receptor sites. Nausea and vomiting may occur with these medications, due to excessive levels of serotonin in the brain. This is a direct result of the vagus nerve receiving too much stimulation from excessive levels of serotonin. Diarrhea may also be a problem, as peripheral serotonin in the digestive tract stimulates intestinal motility.

Another consideration for serotonin psychoactivity stems from the fact that it can serve as a precursor in the healthy human for the production of 5-methoxy-N,N-dimethyltryptamine (5-MeO-DMT), a powerful psychotropic agent, and 5-hydroxy-DMT (bufotenine). From another metabolic route N,N-dimethyltryptamine (DMT) can be produced from tryptophan following the formation of endogenous tryptamine (Figure 3, see Callaway 1995a for a review). Bufotenine has been detected in human urine, particularly after MAO inhibition (Kärkkäinen and Räisänen 1992), which could explain at least some of the psychoactivity occasionally associated with MAO inhibitors. However, bufotenine is not readily psychoactive as a xenobiotic. Because of its polarity, it is not thought to even cross into the brain from the blood. Any psychoactivity from exogenous bufotenine is probably due to its conversion to 5-methoxy-DMT in the body (Callaway et al. 1995).

The purpose of this brief discourse on neurochemistry is to bridge the relationship between the neurochemistry of ayahuasca and the human CNS. There are some striking parallels to consider and a basic familiarity with serotonergic activity is pivotal in understanding the neuropharmacologic actions of this unique beverage.

HARMALA ALKALOIDS FROM
BANISTERIOPSIS CAAPI

The beverage ayahuasca is also known by many other names throughout the Amazon and Orinoco River Basins in northern South America. It is, without a doubt, one of the most sophisticated and complex drug delivery systems in existence (Holmstedt and Lindgren 1967; Callaway 1994a, 1994b; Callaway et al. 1996). The salient feature of this brew is the presence of harmala alkaloids, typically derived from the woody liana *Banisteriopsis caapi,* which serve as inhibitors of the enzyme MAO. The harmala alkaloids, primarily harmine and tetrahydroharmine (THH), are infused from the pounded woody portions of the vine. Methods of preparation vary from simply soaking the pounded material in water overnight to vigorously boiling it for several hours and concentrating the extract.

Most samples of this beverage contain high levels of harmine and tetrahydroharmine (THH), and lesser amounts of harmaline, harmalol, harmol, and related alkaloids (Figure 4, page 102). The primary action of harmine is to temporarily inhibit the activity of MAO (Udenfriend et al. 1958; Buckholtz and Boggan 1977). This action is reversible for the harmala alkaloids, meaning that the enzyme returns to its original state after the inhibiting molecule has been removed by other metabolic processes. For a typical dose of ayahuasca, this translates to approximately 8–12 hours. The inhibitory effect on MAO also depends on the total amount of alkaloid consumed, in addition to the body weight and metabolic rate of the individual.

Harmaline has actions similar to harmine's. It is slightly more potent in its ability to inhibit MAO, although harmaline is not highly concentrated in *B. caapi.* Harmaline may bind to tryptamine sites within the CNS to induce a fine muscular tremor of about 8–12 hertz, which occasionally manifests as nystagmus (Airaksinen et al. 1987; Rommelspacher and Brüning 1984).

THH is another major alkaloid in *B. caapi.* I have also detected this alkaloid in the leaves of *Calliandra pentandra,* which are sometimes added to ayahuasca by the Shuar in Equador, where it is reported to enhance the visionary effects of this brew (Fericgla 1996). This is a bit surprising, as THH is not remarkably psychoactive on its own, even after MAO inhibition (unpublished results). THH, like other 1-methyltetrahydro-ß-carbolines, probably serves to further increase serotonin concentrations by weakly inhibiting serotonin's reuptake into presynaptic neurons after MAO inhibition by harmine (Airaksinen 1980). This effect is not trivial, as the reuptake mechanism and MAO metabolism

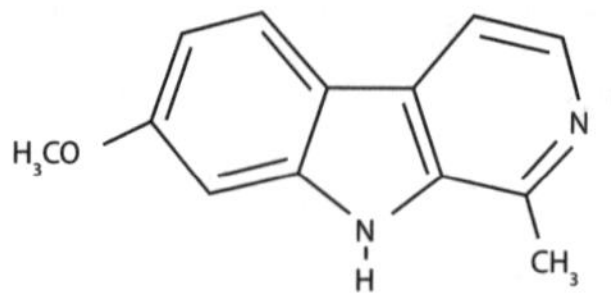

Harmine

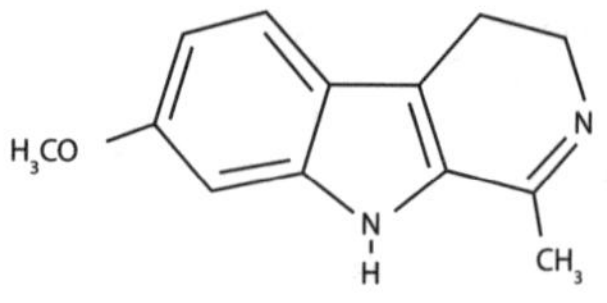

Harmaline

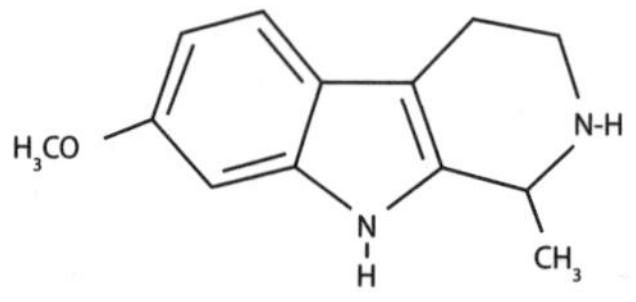

Tetrahydroharmine
THH

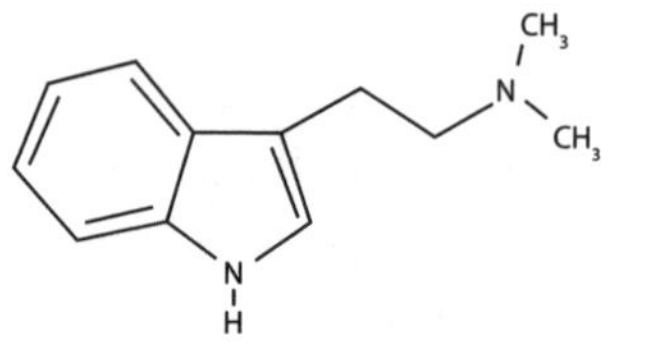

N, N-Dimethyltryptamine
DMT

Figure 4. Three harmala alkaloids typically found in ayahuasca and DMT (note: DMT is not found in *Banisteriopsis caapi* but is an important alkaloid that is often found in ayahuasca).

are the primary processes that clear excess serotonin from the synaptic cleft. It is suggested herein that both MAO inhibition and serotonin uptake inhibition work together through ayahuasca to safely increase levels of serotonin by simultaneously inhibiting both its metabolism and neuronal reuptake, respectively.

Thus, it can be said that a beverage made entirely from *B. caapi* is psychoactive, subsequent to its neuroactivity, although this activity is primarily due to hyperserotonergic actions. Such preparations are used and do have utility, particularly as a purgative and perhaps in removing intestinal parasites. The nausea and vomiting that often occur after drinking ayahuasca are probably due to increasing levels of serotonin in the brain, which results in excessive stimulation of the vagus nerve as previously mentioned. For religious purposes and the induction of visions, however, DMT is typically added to the brew through the leaves of *Psychotria viridis*. This particular combination of plants is nothing short of a blessing from nature.

A young specimen of *Psychotria viridis* under cultivation in Brazil
(photo by Ralph Metzner)

THE SEROTONIN SYNDROME

The serotonin syndrome is a toxic state that results from increased serotonergic activity. It typically results from the combination of MAO inhibitors and some other serotonergic drug, particularly one of the specific serotonin reuptake inhibitors (SSRIs) that are currently so popular in the treatment of depression, obsessive-compulsive disorders, and some forms of substance abuse. This syndrome can even occur by combining tryptophan with a SSRI. In any case, levels of serotonin rapidly increase after its primary metabolic pathways are blocked, i.e., after oxidation by MAO, and its recycling by neuronal uptake into presynaptic neurons.

The serotonin syndrome usually occurs within two hours after dosing, and symptoms may subside within 6–24 hours. The combination of MAO inhibitors with SSRIs is potentially lethal. Initial symptoms include nausea and vomiting, tremor, elevated temperature, cardiac arrhythmia, renal failure, and coma, eventually leading to death. In general, *ayahuasca* should not be used by any individual who is already taking antidepressant medication or other serotonergic drugs, because of its inhibiting action on MAO (Callaway 1993, 1994a; Callaway and Grob 1998).

Brain serotonin levels do not approach toxic levels after ayahuasca alone. While THH may have some capacity to inhibit the reuptake of serotonin, its affinity for this site is apparently much weaker than the potent SSRIs. It is likely that serotonin can even displace THH from this site once serotonin levels have exceeded a certain threshold. It has been suggested that because harmine and harmaline are natural and reversible inhibitors of MAO that such a warning may be premature or even inappropriate. However, the harmala alkaloids are only reversible after several hours, and the lethal effects of the serotonin syndrome have been observed in less time than this (Neuvonen et al. 1993).

Another point of confusion should also be addressed. Occasionally someone who has used ayahuasca on a regular basis for many years begins to take a SSRI, which has been prescribed by a medical doctor. As some tolerance to the serotonergic effects of ayahuasca may have developed over time (Callaway et al. 1994), little or no adverse effects are noted in this case. This is not, however, the same situation as an individual who is already taking a SSRI, and then decides to try ayahuasca for the first time. Furthermore, some SSRIs have active metabolites that remain in the body for several weeks. Thus, it would be wise to allow at least eight weeks to pass from the time of the last dose of a SSRI before ingesting ayahuasca or any other MAO inhibitor.

ADMIXTURES TO AYAHUASCA

It may be safe to assume that the determined hominid has, over time, combined every available plant with B. caapi for a variety of effects. This has resulted in a modification of the typical effects from coca, coffee, tobacco, and various tropane alkaloids from the Solanacea family, to name just a few. It is unlikely that MAO inhibition imparts significantly novel activity to these admixtures. Such alkaloids, unlike DMT, are already orally active on their own. Resulting modifications of native activity would come in the wake of increased serotonin levels in the brain. Such a course of systematic experimentation would eventually lead to the discovery of *Psychotria viridis* as the quintessential ally in ayahuasca technology.

N,N-DIMETHYLTRYPTAMINE (DMT)

DMT is a powerful psychedelic agent (Szára 1956). It also has been used as a cherished molecular tool in making ayahuasca and other sacred products from South America, such as the beverage known as *vino de jurema*, which is traditionally prepared entirely from the root bark of *Mimosa hostilis* (Holmstedt 1995). DMT fits rather well into certain subsets of serotonin receptor sites within the brain (Callaway and McKenna 1998), where it is believed to modify the flow of neuronal information. It has been identified as a natural component of the healthy mammalian brain (Callaway 1995a) and is found in many plant species (Ott 1994; Shulgin and Shulgin 1997). Although a function for its presence in the brain has not been demonstrated, the production of visions in dream sleep has been suggested as a role for endogenous DMT (Callaway 1988).

Unlike most other plants that produce this curious molecule (Ott 1994), *P. viridis* produces DMT almost to the exclusion of any other alkaloid (Figure 4; note the molecular similarity between DMT, psilocin, bufotenine, and serotonin). If one were to assign priority to the many psychoactive molecules that may appear through admixtures in ayahuasca, DMT is, by far, the most dramatic in terms of precipitating visual phenomenon. It is a relatively simple molecule that is not orally active, despite its similarity in effect and molecular structure to psilocin (Figure 1), which is orally active.

Under ordinary circumstances DMT is rapidly metabolized by MAO, the same enzyme that metabolizes serotonin. With MAO intact, the potential activity of DMT might be overlooked by the grazing biped. After MAO is inhibited by harmala alkaloids, for example, DMT becomes orally active and intricate visual displays of colored patterns are often achieved through this combination. It is worth noting that the effects from orally activated DMT are qualitatively different from either injected or smoked DMT in the absence of MAO inhibition (Strassman et al. 1994; Callaway et al. 1998, respectively). The onset of effects begins within one minute and lasts up to five or ten minutes after injecting or smoking DMT. The interval between this time is typically filled with a fleeting display of color and geometry. Full-blown visions and complete dissociation from physical reality is not unusual at higher dosages. After MAO inhibition, however, the onset of oral DMT begins after twenty minutes and typically lasts up to ninety minutes, and the psychic effects tend to be integrated while awareness of the physical environment is typically intact.

This difference is partly due to the route of administration of DMT, the amount of time involved, and almost certainly due to the influence of increased serotonin in the brain after MAO inhibition. The latter point is important because in this case DMT must compete with serotonin for receptor binding sites to have an effect. Essentially, in this regard, serotonin serves to keep the trip on track. The trade-off for DMT between intensity (i.e., injected or smoked) and time (lengthened after MAO inhibition) allows for more cognitive functions to engage the experience in the later scenario. When smoked or injected, DMT seems to have all the cognitive content of a fireworks display.

AYAHUASCA IN CONTEMPORARY USAGE

Aside from its indigenous use, ayahuasca has been embraced by those residing in urban populations as well. In addition, purely synthetic preparations have been described, which are based on this ancient technology (Callaway 1993, 1995b; Ott 1994). While many plants are often used in the traditional brews, modern usage seems limited to beverages prepared exclusively from *B. caapi* and *P. viridis*. Approximately ten thousand regular participants take part in these new religions, typically twice a month and sometimes once a week. In traditional usage, only a small percentage of the population uses ayahuasca on such a regular basis, although almost everyone in the society will have had the experience at least three to five times in life, at various stages of physical and personal development (Fericgla 1997). In contemporary usage, however, all adult members are typically encouraged to participate at every opportunity. Adolescents do not participate on a regular basis until well into their teens, although there are exceptions.

Of the contemporary groups, much variation is readily apparent. The União do Vegetal (UDV), the largest unified ayahuasca church, sit throughout their sessions, while members of the Santo Daime and Barquinia typically dance. The Santo Daime dance an intricate two-step, where men and women are ordered into discrete configurations, while followers of Barquinia ambulate in a snaking line. The UDV structure is essentially democratic, while the Santo Daime remains a paternal hierarchy, and the Barquinia are led by a charismatic Black woman. These are broad generalizations to illustrate the cultural variety that can exist within the usage of ayahuasca.

The quest for this particular effect, and the fact that it is imparted by known chemical entities, is highlighted by the following example. An ayahuasca church decided that a responsible practice would be the cultivation of their own *B. caapi* and *P. viridis,* rather than constantly returning to the forest to harvest the raw material. After several years, the plants were ready for use. Unfortunately, some of the desired effects were not apparent. Many hypotheses were offered over time, some physical and some metaphysical. After careful botanical inspection it was eventually realized that *P. nervosa* was planted rather than *P. viridis*. The former species of *Psychotria* is essentially devoid of DMT although the physical features of these plants are almost identical. As an interesting aside, the specimens of *P. nervosa* were left to grow as decorative varieties around the temple, perhaps as a reminder of this lesson from nature.

THE HOASCA PROJECT

In 1992, Dennis McKenna began to collect people for a prospective study of an ayahuasca beverage used by the UDV in Brazil, where it is called hoasca. Some time before 1992, the UDV had already decided to contact Dr. McKenna to initiate this study. Hoasca is made exclusively from *Banisteriopsis caapi* and *Psychotria viridis,* which helped to define the tea in a botanical sense, and all phases of its preparation were documented and measured. Also, the practice of sitting throughout the acute hoasca session was especially convenient for measuring anatomical responses and collecting blood samples on a rigid time schedule.

Two varieties of *Banisteriopsis caapi;* two pieces each of *tucunaca,* on the right, and *caupurí* on the left. Note the smooth cylindrical features of the *tucunaca* as compared with the knobby appearance of *caupurí.* (photo by J. C. Callaway)

In all, fifteen male volunteers participated in the clinical phase of this study. All gave informed consent and were in good health, and all had used the tea on a regular basis for at least ten years. Some of the more interesting phytochemical and pharmacological findings from this study are presented in the following text.

VARIATIONS OF HARMALA ALKALOIDS IN SAMPLES OF *B. CAAPI*

There seems to be at least two varieties of *B. caapi* that are both chemically and morphologically distinct. The UDV refer to these two as *tucunaca* and *caupurí.* The tucunaca is a smooth vine, while the caupurí has large internodes where the stems are produced. The tucunaca grows in the cooler climes of southern Brazil and is known for its milder effects, especially on the digestive tract. The caupurí grows in the hot and steamy regions of northern Brazil and is known for its strong purgative effects. A recent phytochemical survey of thirty-three *B. caapi* samples collected on the same day throughout Brazil has been published (Callaway, Brito, and Neves 2005).

When preparing the tea, according to the UDV, a standard methodology is used. While the actual amounts of added plant material are up to the mestre's discretion, the physical proportions of *B. caapi* and *P. viridis* remain essentially the same. In theory, larger amounts of tucunaca could be used to make a tea similar to caupurí, or vice versa. In practice, however, the result is a tea containing lesser amounts of harmala alkaloids from the tucunaca variety of *B. caapi*.

Another item of phytochemical interest is the remarkable change in harmala profiles over various samples of *B. caapi*, which seem to be similar in apparent morphology. In some of the thirty-five samples of *B. caapi* analyzed, fairly high amounts of THH and/or harmine are present, which seems to have a direct impact on the overall quality of the tea. In particular, experienced drinkers seem to prefer those teas where THH concentrations were high, relative to harmine and harmaline. They explained that such teas delivered more "force" to the experience. Overall, harmine concentrations were found to be highest in the analytical analysis. This was followed closely by THH, then lesser amounts of harmaline (unpublished results).

VARIATIONS OF DMT IN SAMPLES OF *P. VIRIDIS*

From thirty-seven samples of *P. viridis,* collected in the morning on the same day from several locations throughout Brazil, the DMT concentrations were found to range from 0.00–17.75 milligram per gram (mg/g) of dried leaf (Callaway, Brito, and Neves 2005). Most samples had a value of approximately 7.50 mg/g DMT, and only one had undetectable amounts. It is possible that this specimen was not *P. viridis,* but another species of *Psychotria.*

Most remarkable were the results from consecutive samples taken from the same plant at several different times throughout the day. The highest levels of DMT were found in those leaves that were collected at dawn (8.97 mg/g) or before dusk (9.52 mg/g DMT). The leaves that had the least amount of DMT were those collected at midnight (5.57 mg/g), and another depression in alkaloid content appeared near 10:00 AM (8.01 mg/g DMT) and subsequent values remained low throughout the hotter parts of the day. One could argue that these are simply variable values from different leaves, although special consideration was taken to include leaves at equivalent stages of development. It is also important to note that these quantitative results are in agreement with what has been observed qualitatively over years of practical experience.

VARIATIONS OF ALKALOID CONCENTRATIONS IN CONTEMPORARY AYAHUASCA

As one might expect, variability in alkaloid profiles within a particular species are compounded by adding additional plants to the brew. Thus, teas prepared from these two plants offer unique and complex mixtures (Callaway 2005a). Of the teas that were offered by the UDV, a total of twenty samples were analyzed for alkaloids and these results are presented in Table 1. Such a collection of results cannot say much about individual brews, and even less about a particular individual's experience. One point worth making, however, is that a few teas had no detectable amounts of DMT, yet they were still considered useful in a ceremonial sense. This brings up many interesting questions about the activity of such complex mixtures, not to mention a technical aside that not all forms of ayahuasca necessarily contain DMT, although appropriate plants were used to prepare the beverage.

	DMT	TETRAHYDRO-HARMINE	HARMALINE	HARMINE
Averages	1.12 ±1.37 mg/ml	1.82 ±1.03 mg/ml	0.23 ±0.26 mg/ml	2.25 ±1.48 mg/ml
Ranges	0.00–5.84	0.45–5.26	0.00–0.90	0.45–6.25

Table 1. Averaged alkaloid contents of twenty teas prepared by the UDV, expressed as milligrams per milliliter of tea (mg/ml), ± the standard deviation, and ranges of minimum and maximum values.

VARIATIONS IN INDIVIDUAL METABOLISM

Perhaps all men are created equal, as the saying goes, but they certainly seem to metabolize hoasca differently. In the Hoasca Project we measured four alkaloids in the blood of fifteen volunteers over a twenty-four-hour period. The hoasca was administered by a mestre, according to their religious practice, at a dosage of 2 milliliters per kilogram of body weight, which was not according to their religious practice. The average amounts and ranges of hoasca, body weight, and alkaloids consumed are presented in Table 2. The intensity of subjective effects from the tea were closely correlated with peak plasma alkaloid concentrations in all cases, especially DMT.

	BODY WEIGHT	TEA	HARMINE	HARMALINE	THH	DMT
Averages	74 kg	148 ml	252 mg	30 mg	159 mg	36 mg
Ranges	58–90	120–180	204–306	24–36	128–193	29–43

Table 2. Average amounts, and ranges, of body weights, amounts of tea ingested, and total amounts of alkaloids consumed by fifteen volunteers during the pharmacokinetic phase of the Hoasca Project. These results are presented in terms of kilograms (kg) body weight, milligrams (mg) of alkaloid, and milliliter (ml) of tea ingested.

Of the fifteen volunteers, one was eliminated from the analysis because he vomited early into the session and seven were found to metabolize harmine significantly faster than the remaining seven volunteers (Callaway 2005b). Surprisingly, DMT levels were not significantly affected by this difference in harmine metabolism. It was also noted that the plasma concentrations for THH were not greatly influenced in this regard, suggesting that its metabolism may not be significantly tied to MAO inhibition.

These findings have little bearing on how ayahuasca is actually used, however, as the shaman or mestre is ultimately responsible for the amount to be given. This is a judgment often based on years of experience, the strength of the tea, and some knowledge of the person(s) receiving the tea.

PHYSIOLOGICAL CHANGES AFTER THE INGESTION OF HOASCA

Neuroendocrine responses all increased well over basal levels for each volunteer, with maximum concentrations for cortisol, growth hormone, and prolactin being achieved by 60, 90, and 120 minutes respectively. All of these measures were back down to baseline by 360 minutes. Such a sharp response is not unusual for other serotonergic drugs (e.g., MAO inhibitors).

Anatomical measurements were also taken. Pupil diameter increased, along with the subjective effects, from 3.7 mm after 40 minutes to a maximum of 4.8 mm after 180 minutes. Pupils remained dilated after the last measurement at 240 minutes and appeared normal after 6 hours. Respiration increased slightly to a maximum of 22 breaths per minute at 90 minutes, then fluctuated as the session progressed. Oral temperature increased slightly, but not to a significant degree.

Cardiovascular measures initially increased over basal levels for each volunteer. Maximum heart rate (79 beats per minute) was recorded after 20 minutes, which dropped below basal levels after 120 minutes, to a minimum of 65 bpm, and gradually began to approach basal levels after 240 minutes. Both systolic and diastolic pressure increased to 137 and 92 mm Hg, respectively, after 40 minutes, then gradually returned to basal levels after 180 minutes. At 240 minutes, however, the systolic and diastolic pressures had dropped to 124 and 81 mm Hg, respectively. While these values may be considered low, they can also be seen as resembling those induced by a deep meditative state, which is essentially what was observed and apparently experienced. It seems likely that higher cardiac values would have been observed had these volunteers been dancing instead of sitting.

CONCLUSIONS

Psychoactive agents are typically not consumed for their food value, and archaeological evidence suggests their use precedes any written record. Such a long and, in some cases, continuous history strongly suggests a utility for the use of psychoactive substances as adjuncts to religious practice.

If ayahuasca is not the most complex binary drug delivery system in existence, what is? It certainly must be the oldest. Exactly how the technology was devised to locate and combine certain plants to enable the oral activity of DMT remains a mystery. The only certainty in this area is in the identification of active plant alkaloids and their subsequent effects on human consciousness.

References

Airaksinen, M. M., H. Svensk, J. Tuomisto, and H. Komulainen. 1980. Tetrahydro ß-carbolines and corresponding tryptamines: In vivo inhibition of serotonin and dopamine uptake by human blood platelets. *Acta Pharmacologia et Toxicologia* 46:308–13.

Airaksinen, M. M., A. Lecklin, V. Saano, L. Tuomisto, and J. Gynther. 1987. Tremorigenic effects and inhibition of tryptamines and serotonin receptor binding by ß-carbolines. *Pharmacology & Toxicology* 60:5–8.

Buckholtz, N. S., and W. O. Boggan. 1977. Monoamine oxidase inhibition in brain and liver by ß-carbolines: Structure-activity relationships and substrate specificity. *Biochemical Pharmacology* 26:1991–96.

Callaway, J. C. 1988. A proposed mechanism for the visions of dream sleep. *Medical Hypotheses* 26:119–24.

———. 1993. Tryptamines, ß-carbolines and you. *MAPS Newsletter* 4(2):30–32.

———. 1994a. Another warning about harmala alkaloids and other MAO inhibitors. *MAPS Newsletter* 4 (4): 58.

———. 1994b. Some chemistry and pharmacology of ayahuasca. *Jarbuch für. Ethnomedizin und Bewußtseinsforschung* (Yearbook of Ethnomedicine and the Study of Consciousness) 3:295–98.

———. 1995a. DMTs in the human brain. *Jarbuch für Ethnomedizin und Bewußtseinsforschung* (Yearbook for Ethnomedicine and the Study of Consciousness) 4:45–54.

———. 1995b. Pharmahuasca and contemporary ethnopharmacology. *Curare* 18 (2): 395–98.

———. 2005a. Various alkaloid profiles in decoctions of *Banisteriopsis caapi*. *Journal of Psychoactive Drugs* 37 (2): 151–55.

———. 2005b. Fast and slow metabolizers of *hoasca*. *Journal of Psychoactive Drugs* 37 (2): 157–61.

Callaway, J. C., M. M. Airaksinen, D. J. McKenna, G. S. Brito, and C. S. Grob. 1994. Platelet serotonin uptake sites increased in drinkers of *ayahuasca*. *Psychopharmacology* 116:385–87.

Callaway, J. C., M. M. Airaksinen, and J. Gynther. 1995. Endogenous ß-carbolines and other indole alkaloids in mammals. *Integration—Journal of Mind Moving Plants and Culture* 5:19–33.

Callaway, J. C., L. P. Raymon, W. L. Hearn, D. J. McKenna, C. S. Grob, and G. S. Brito. 1996. Quantitation of *N,N*-dimethyltryptamine and harmala alkaloids in human plasma after oral dosing with *Ayahuasca*. *Journal of Analytical Toxicology* 20:492–97.

Callaway, J. C., and C. S. Grob. 1998. *Ayahuasca* preparations and serotonin re-uptake inhibitors: A potential combination for severe adverse interaction. *Journal of Psychoactive Drugs* 30:367–69.

Callaway, J. C., and D. J. McKenna. 1998. Neurochemistry of psychedelic drugs. In *Drug Abuse Handbook*, chapter 6.6, ed. S. B. Karch, 485–98. Boca Raton: CRC Press.

Callaway, J. C., D. J. McKenna, C. S. Grob, G. S. Brito, L. P. Raymon, R. E. Poland, E. N. Andrade, and E. O. Andrade. 1999. Pharmacokinetics of *Hoasca* alkaloids in healthy humans. *Journal of Ethnopharmacology* 65 (3): 243–56.

Callaway, J. C., G. S. Brito, and E. S. Neves. 2005. Phytochemical analyses of *Banisteriopsis caapi* and *Psychotria viridis*. *Journal of Psychoactive Drugs* 37 (2): 145–50.

Fericgla, J. M. 1997. *Al Trasluz de la Ayahuasca: Coleccion Cogniciones, Estados Modificados de Consciencia*. Barcelona: Los Libros de la Liebre de Marzo.

Grob, C. S., D. J. McKenna, J. C. Callaway, G. S. Brito, E. S. Neves, G. Oberlander, O. L. Saide, E. Labigalini, C. Tacla, C. T. Miranda, R. J. Strassman, and K. B. Boone. 1996. Human psychopharmacology of Hoasca, a plant hallucinogen used in ritual context in Brasil. *Journal of Nervous and Mental Disease* 184 (2): 86–94.

Heffter, A. 1896. Über Cacteenalkaloide. II Mittheilung. *Berichte der Deutschen Chemischen Gesellschaft* 29:216–27.

———. 1898. Über pellote. Beiträge zur chemischen und pharmakologischen kenntnis der cacteen. Zweite Mittheilung. *Archiv für Experimentelle Pathologie und Pharmakologie* 40:385–429.

Hofmann, A. 1978. History of the basic chemical investigations on the sacred mushroom in Mexico. In *Teonenácatl: Hallucinogenic Mushrooms of North America, Psycho-Mycological Studies* No. 2, eds. J. Ott and J. Bigwood. Seattle, WA: Madrona Publishers.

Holmstedt, B. R. 1995. Personal communication.

Holmstedt, B. R., and J. E. Lindgren. 1967. Chemical constituents and pharmacology of South American snuffs. In *Ethnopharmacologic Search for Psychoactive Drugs*, eds. D. H. Efron, B. Holmstedt, and N. S. Kline. U.S. Public Health Service Publication No. 1645. Washington, D.C.: GPO.

Kärkkäinen, J., and M. Räisänen. 1992. Nialamide, an MAO inhibitor, increases urinary excretion of endogenously produced bufotenin [*sic*] in man. *Biological Psychiatry* 32:1042–48.

Luna, L. E., and P. Amaringo. 1991. *Ayahuasca Visions: The Religious Iconography of a Peruvian Shaman*. Berkeley, CA: North Atlantic Books.

McKenna, D. J. 1992. Personal communications.

McKenna, D. J., and G. H. N. Towers. 1984. Biochemistry and pharmacology of tryptamines and beta-carbolines: A minireview. *Journal of Psychoactive Drugs* 16 (4): 347–58.

McKenna, D. J., G. H. N. Towers, and F. Abbot. 1984a. Monoamine oxidase inhibitors in South American hallucinogenic plants: Tryptamine and ß-carboline constituents of *ayahuasca*. *Journal of Ethnopharmacology* 10 (2): 195–223.

McKenna, D. J., G. H. N. Towers, and F. Abbot. 1984b. Monoamine oxidase inhibitors in South American hallucinogenic plants. Part 2: Constituents of orally-active myristicaceous hallucinogens. *Journal of Ethnopharmacology* 12 (2): 179–211.

Neuvonen, P. J., S. Pohjola-Sintonen, U. Tacke, and E. Vuori. 1993. Five fatal cases of serotonin syndrome after moclobemide-citalopram or moclobemide-clomipramine overdoses. *Lancet* 324:1419.

Ott, J. 1994. *Ayahuasca Analogues: Pangæan Entheogens*. Kennewick, WA: Natural Products Co.

Rommelspacher, H., and G. Brüning. 1984. Formation and function of tetrahydro-ß-carbolines with special reference to their action on [3H]tryptamine binding sites. In *Tryptophan*, eds. H. G. Schloßberger, H. Kochen, A. Linzen, and R. Steinhart. Berlin: DeGruyter.

Schultes, R. E., and A. Hofmann. 1992. *Plants of the Gods: Their Sacred, Healing and Hallucinogenic Powers*. Rochester, VT: Healing Arts Press.

Schultes, R. E., and R. F. Raffauf. 1992. *Vine of the Soul: Medicine Men, Their Plants and Rituals in the Columbian Amazon*. Oracle, AZ: Synergistic Press.

Shulgin, A., and A. Shulgin. 1997. *TIHKAL: The Continuation*, ed. Dan Joy. Berkeley, CA: Transform Press.

Spruce, R. [1908] 1970. *Notes of a Botanist on the Amazon and Andes* (two volumes), ed. A. R. Wallace. London: Macmillan. Reprint, New York: Johnson Reprint.

Strassman, R. J., and C. R. Qualls. 1994. Dose-response study of N,N-dimethyltryptamine in humans I. Neuroendocrine, autonomic, and cardiovascular effects. *Archives of General Psychiatry* 51:85–97.

Strassman, R. J., C. R. Qualls, E. H. Uhlenhuth, and R. Kellner. 1994. Dose-response study of N,N-dimethyltryptamine in humans II. Subjective effects and preliminary results of a new rating scale. *Archives of General Psychiatry* 51:98–108.

Szára, S. 1956. Dimethyltryptamin [*sic*]: Its metabolism in man; the relation of its psychotic effect on serotonin metabolism. *Experientia* 12:441–42.

Udenfriend, S., B. Witkop, B. Redfield, and H. Weissbach. 1958. Studies with the reversible inhibitors of monoamine oxidase: Harmaline and related compounds. *Biochemical Pharmacology* 1:160–65.

4

THE EXPERIENCE OF
AYAHUASCA

TEACHINGS OF THE AMAZONIAN
PLANT SPIRITS

The following twenty-four personal accounts describe the ayahuasca experiences of people from a wide variety of backgrounds—most, though not all, from North America or Europe. Most of the sessions fall under the category that I'm calling hybrid shamanic psychological rituals, although a few describe experiences with one or the other of the Brazilian churches that use the vine. In their own words, these individuals describe the healing insights, perceptions, and emotional responses to the vision-teachings they received in the ayahuasca state. Since, in many cases, quite some time had elapsed from the time the sessions occurred and the time these accounts were written, the contributors were also asked to reflect on the long-term consequences and significance of their ayahuasca journeys.

In choosing these personal accounts, I was particularly interested in selecting those who reported more ecological consciousness, more awareness of the threats and challenges facing our culture, and a greater commitment to a lifestyle that incorporates both spiritual practice and respect for all of nature. This is not to deny that some ayahuasca experiences may be painful and unpleasant and that some may not have very profound consequences, but to reinforce the point that we are not dealing primarily with the automatic effects of a drug; rather we are dealing with an ancient shamanic initiatory practice in which the vine-tea can function to amplify one's awareness of the interconnected web of all life.

RALPH METZNER

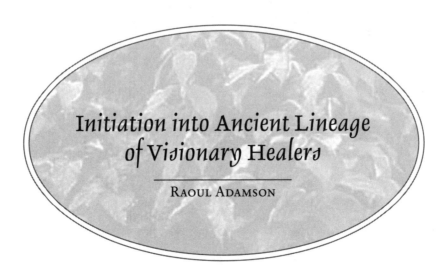

Initiation into Ancient Lineage of Visionary Healers

RAOUL ADAMSON

The following account, by a psychologist in his fifties, illustrates many of the classic elements of ayahuasca visions: the encounters with snakes and jaguars, the sense of being met, taught, and healed by conscious intelligent beings or spirits. He learns to sing, in self-defense against abusive behavior. There are insights into the processes of perception and judgment and a profound feeling of acceptance and gratitude.

My initiation to ayahuasca occurred by way of an ethnobotanist friend who had spent considerable time in South America studying with mestizo *ayahuasqueros* in Peru. He had learned how to grow the two plants that compose the medicine in Hawaii and had prepared the brew according to the traditional recipe. The setting was a spacious house set among trees in a rural area of northern California. We drank the brew, which has a taste that is a strange mixture of bitterness and syrupy sweetness, in almost total darkness, with only a candle or two. We listened to the Mayan music of *Xochimoci*. I began to feel very relaxed, heavy, and soft, but also as if my head were expanding.

A swaying tapestry of visions came into view, at first mostly geometric patterns, which looked familiar from previous experiences with tryptamine hallucinogens, including psilocybe mushrooms. As usual, I experienced these geometric patterns with distaste verging on disgust: they seem tacky, plastic, and artificial, like the décor of a shopping mall or a Las Vegas casino. As I searched for the meaning of my reaction, I was shown how this is the human technocultural overlay on the natural world: I was looking at the human world! As I accepted that, with some regret, I was able to see through it to the pulsating energies of the world of all-encompassing nature, permeated by spiritual, astral beings and forms.

There were shapes and images of plants, animals, humans, ethereal temples and cities, flying craft, and floating structures. Particular images from time to time emerged out of the continuous flux and then were reabsorbed back into it. As the images of forms and objects receded back into the swaying fabric of visions, I realized that I was seeing them as if projected on the twisting coils of an enormous serpent with glittering silvery and green designs on its skin. I could not see either head or tail of the serpent, which gave me a rough sense of its size: it encompassed the entire two-story building. Curiously, the sight of this gigantic serpent did not evoke the slightest fear; on the contrary, my emotional response was one of awe and humility at the magnificence of this being and its spiritual power. I was reminded of Pablo Amaringo's ayahuasca paintings, which depict this giant serpent seen in the visions as the "mother spirit," on which other smaller spirits ride and travel. In the Amazon region they see three different serpent mamas—of the air, of the river, and of the forest. Here there seemed to be one enormous serpent mother, coiling and rippling through the entire length and breadth of the valley in which we are situated.

Then I met another serpent in my visions, more "normal" in its dimensions: in fact it was about the same size as me. It entered my body through my mouth and started to slowly wind its way through my stomach and intestines over the next two or three hours. When it got to the gut, there was some cramping, and incredibly loud sounds of gurgling and digesting were coming from my viscera. I became aware of a morphic resonance between serpent and intestines: the form of the snake is more or less a long intestinal tract, with a head and a tail end; and conversely, our gut is serpentine, with its twists and turns and its peristaltic movement. So the serpent, in winding its way through my intestinal tract was "teaching" my intestines how to be more powerful and effective—certainly a gut-level experience!

Then I saw several black-skinned people, dancing as they came toward me and receded away. They were always in pairs, like twins, moving in parallel fashion: I wondered whether they represented the spirits of the two paired plants of the ayahuasca tea. Then, as I was lying sideways on a couch, a jaguar suddenly came into me. It was an enormous black feline male, and he entered my body assuming the same semireclining position I was in. Shortly after I noticed it, the jaguar was gone. Another time, as I was on my hands and knees, I distinctly felt a bird landing on my back. I was being briefly introduced to some of the different spirits that the ayahuasca medicine can access. The realization grew within me that with practice and increased concentration, I would be able to hold the encounters with the different animal spirits for longer, and then be able to question them for divination. Don Fidel, one of the old ayahuasqueros, said: "the visions come into you and heal you."

Images of Mayan gods and underworld demons dancing appeared: skeletal, crippled, diseased, skin flapping, blood dripping, pustular, bulbous, with gaping wounds and cut-off heads, toads on their necks, pierced with thorns. Their message, repeated several times, was: "*you* don't have to *do* anything." By incorporating death, decay, disease, and other unimaginable horrors into their dance of transformation, a deep inner healing took place, seemingly independent of any personal involvement on my part. I was astonished at being initiated into this ancient lineage of visionary healers.

It was late in the evening, and I was again on my hands and knees, feeling overwhelmed and exhausted by this gut-wrenching, yet soul-refreshing, journey through the netherworlds of jungle, river, and serpents. I lowered my forehead to touch the ground, then I realized I was falling slowly through the earth, through soil and rock, moving faster and faster, and then dropping out the other side into deep space, vast in its darkness, exhilarated, filled with countless points of light, *scintillae,* luminous streaks and stars of the universe.

My next encounter with the vine of visions occurred on New Year's Eve. I wanted to explore the experience of a larger amount of the medicine and asked my partner to "sit" for me. We were in a house by the ocean and had tape recordings of *icaros,* the healing chants of the ayahuasqueros, made by Eduardo Luna. In the days before I had been thinking about sacrifice and self-sacrifice, and wondering what the experience of being eaten by another animal was like. We humans, having become top predator in the food chain, have not had that experience for many thousands of years. But we used to: our existence during the Stone Age and Ice Age was surely marked by many fatal encounters with large predators. I had a lingering concern that this relationship between two organisms as "eater" and "food" was somehow imbalanced. Given that all processes of life work on the principle of perpetual balance and exchange, what, I wondered, did the one who is eaten get out of it? The ayahuasca spirits answered my question as soon as I posed it: "If I am eaten by the serpent, I acquire its power and knowledge. Allow someone to eat you and you gain their power." Suddenly I thought about the Mayan sculptures and paintings showing a gigantic serpent-dragon, with the human face of a god looking out of its jaws. I felt ready to be eaten.

In my first experience with ayahuasca, I had "swallowed" the serpent—although it certainly seemed as if the serpent was taking the initiative. This time, it was my turn to be swallowed, and digested, by the serpent. When the visions started they were connected with various involuntary bodily reactions, especially excretion, purging, burping, and farting. The alchemical furnace was cooking, and various gases and fluids were spraying and oozing out, on an etheric or psychic level mostly, and occasionally on the physical level as well. It wasn't painful, just strange and unusual. I felt like I was *in* areas of my body I had never consciously been in before, this lifetime. At one point I felt I was turned inside out—some force reached into my mouth and throat and pulled my insides out, until my inner organs were all on the outside, hanging out, so to speak, and limbs and muscles had become packed inside.

The ayahuasca jungle elves, the little green guys, were carrying away what looked like armor plates and metallic pieces. I got the sense they were taking apart pieces of a structure, to wash and polish them and tune them up for better functioning. Suddenly I realized the structure they were dismantling was myself. I yelled after them (inwardly), "Hey, wait, that's *me* you're carrying away there." Without missing a beat, they replied cheerfully, "Not to worry, we'll put you back together, you'll be fine." All the time they were singing in the rhythmic chants of the icaros we were hearing. I had experienced, and heard of, shamanic dismemberment experiences before, where you are pounded and pulverized, or sliced and cut up, as a prelude to eventual healing reconstitution. But this was the first time I experienced this kind of civilized, courteous, efficient dismantling. The green elves were taking apart my character armor, and giving me back an improved, more flexible, more comfortable body-mind.

The icaros provided an essential support for this kind of radical transformation of somatic consciousness. Without them I'm sure I would have felt lost and frightened. The rapid rhythm of the chants kept me moving through the jungle of visions. The warm, deep-throated voice of the ayahuasqueros brought soothing comfort, making even the most outrageously intimate interventions of the medicine tolerable. At one point, I found myself sinking into a mood of profound grief and anguish, being wracked by heart-wrenching sobs. But I could not tell what this grief was about. Then I realized I had been listening to a portion of the recording in which a woman being healed in an ayahuasca ceremony is telling her story. I could not understand the words, but they were accompanied by anguished sobbing and crying. It seemed as if I was experiencing the feelings of someone relating the loss of a loved one. After her anguished lament, the warm, soothing voice of the ayahuasquero, with infinite tenderness, offered a healing balm for the tormented soul. I later found out that, in fact, that track on the recording was the healing of a woman whose husband had been killed.

I had two moments of considerable anxiety. The first occurred when I was experiencing so many extremely unfamiliar physiological sensations, such as being turned inside out, that I couldn't tell whether my vital signs, i.e., pulse and breathing, were okay. So I asked my partner to verify that I was indeed exhibiting all the normal signs of healthy aliveness. After receiving reassurance on that score, I then started to worry that I might be losing my mind, going really insane. Again I asked my partner, and after a moment of consideration she replied that since I posed the question in a perfectly rational manner, it did not seem to her that I was showing signs of insanity. From this I concluded that the rational and reflective capacities are not impaired during these interior journeys; it's just that an enormous variety of nonrational, hyper-sensory, and totally novel perceptions are added to the total stream of experience.

I was given teachings concerning the attitude and actions of the warrior in relation to the visions. The visionary warrior is not just passively taking in the visions, as we do when watching a film or television, or during most dreaming. The warrior is actively looking at them, observing the details, searching for the meaning behind the appearances. This is what is meant by the warrrior's impeccability: s/he is without stain or blemish, no egocentric projections are distorting perception. The warrior is never a victim and does not idealize the condition of vulnerability. Although s/he can be wounded of course, as anyone can, whether in the jungle of subjective visions or the urban jungle, such wounds are taken care of in the appropriate way.

Related to this teaching was a method of analyzing perception that I had learned in an earlier visionary state, and that I continued to practice. I was shown that there are three phases in the process of sense perception, whether of inner or outer reality: first, there is experiencing, the pure sensory contact; second, there is the action of the inner observer, witnessing the experience; and third, there is the recording and communicating of the experience, which necessarily involves external expression in word, or sound, or paint, or other medium. Buddhist mindfulness meditation (vipassana), I realized, focuses on the second phase, the witnessing. To really complete an experience, I was shown, we need to go through all three phases. Painful or traumatic experiences are often incomplete, sometimes because there are powerful prohibitions on the third phase of communicating. Healing or recovery from trauma involves telling the story, so that it is shared, believed, and recognized.

In preparing for my next experience with ayahuasca, I asked my guiding spirits to help me bring through some songs (icaros), as I had been very impressed by the power of the songs to channel the visions in a healing direction. The way this came about was a complete surprise. I took the medicine with two other friends, one with considerable experience with Amazonian healing rituals, the other a North American physician-healer, for whom it was the first experience. A woman friend of mine and the wife of the physician were also there as sitters. When the medicine started to come on, the physician expressed his discomfort with the changes he was experiencing by loud and repeated retching into a bowl that he cradled in his lap, and by hurling one insult after another at myself and my other friend. The emotional purging of his hostile outbursts paralleled the physical purging of his vomiting.

At first I thought this was an initial shock reaction that would eventually pass. But as it continued and showed no signs of abating, a feeling of considerable annoyance and irritation came up. My more experienced friend was similarly annoyed and withdrew under a blanket into a cocoon of silence. As I learned later, his fear was that he might be triggered into a similar negative reaction. My own reaction was slightly different: I found myself thinking that I didn't want this barrage of negativity to destroy or distort the positive, healing journey I had envisioned for myself. I laid down, turning away from the others, and suddenly found myself chanting simple rhythmic phrases, in a low, soft voice. Some of them related to the hostile verbiage still coming from our physician friend: "he's telling people what's what, and what's not," or "why is he kicking me." They functioned to ward off the attack energies, which gradually receded into the background of my awareness.

The chant that continued went something like this, with numerous repetitions, different combinations and variations of the lines, in a kind of sing-song rhythm:

"O the woman, O the man / O the man, O the woman / O the song of the woman / O the song of the man / O the song of the longing / O the longing of the song / O the song of the longing / of the woman for the man / O the longing of the song / of the man for the woman." The tone of my voice changed back and forth between masculine and feminine, as the chant emanated a soothing, comforting aura through our little group. Eventually, even our agitated friend calmed down. I was profoundly moved by this gift of healing song from the ayahuasca spirits. I had known that songs could heal before, but what was new to my understanding here was that they could also function in a protective manner against toxic emotional negativity.

A memorable journey with ayahuasca took place in Hawaii, in a house on the slopes of the Mauna Loa volcano on the Big Island. Five men and three women constituted the group, all fairly experienced travelers. There were some differences of expectation that had not been resolved, leading to some uncertainty, at least on my part, on whether singing would be acceptable. One woman was vomiting with volcanic intensity. I decided to pursue whatever was presented to me.

Lying on the floor, looking up at the pattern of light and shade made by the ceiling beams, I had two visions. First, I saw the outline of the body of a huge lizard, looking up at him from below, as if he had just landed on a skylight in the roof. The image stayed there the whole night. I felt he was protecting us. He was Itzamna, the Mayan sky god, who takes the form of a lizard. Later, I became an iguana, feeling the unusual sensation of having a heavy tail, as long again as my whole torso.

The second image on the ceiling was of a man, sitting at the entrance of a cave, looking out into a brighter landscape. He had a pointed hat or cap on and was carrying a large staff. I had no idea what to make of this image; it too persisted, even when I returned from an interaction involving other people and situations. Even with repeated questioning, I could not get any further clarification or elaboration on the basic theme, nor any understanding of the meaning of the vision.

Not until about eight months later, when I was on a fasting vision quest in the White Mountains of California, did I finally see the meaning of that vision. I was sitting at the entrance to a cave where I was going to spend the night, looking down into the valley flooded with evening light, when I suddenly remembered the ayahuasca vision: *I was the man, with a pointed cap and a staff, sitting at the entrance to a cave.* The vision had been precognitive or prophetic. This was one of many visions and dreams that have led me over the past years to the conviction that many of our visions and dreams are partially precognitive, but it is easy to miss the connection if we don't make a point of checking and comparing our recorded perceptions with subsequent events and experiences.

In a final experience I can relate here, I took a faily large amount of ayahuasca alone, accompanied only by my partner. My intention was to gain more insight into the processes of visionary experience. I was taught a number of very valuable lessons, confirming insights from Buddhist and other disciplines of meditation. After one to two hours, the swirling, swaying mass of kaleidoscopic, geometric shapes flowed around and through me, softly exploding and imploding, changing so rapidly that I was unable to verbalize any description. I was learning to shift my focus of attention back and forth between visions, which are more "up there," and bodily sensations, which are more "down there." For short periods I could even hold both simultaneously. When I tried to speak however, they both disappeared.

I noticed that when I started to worry about some bodily sensation, thinking about it in a hypochondriacal way, the pulsing flow of images slowed down and stopped. When I breathed deeply and stopped worrying, it started up again. So my first lesson, confirming what my meditation teachers had taught me, was: *stop worrying* about your experience, it only blocks the flow of energy.

Then I also noticed that when I judged any aspect of my visions as being ugly or horrible or bad, this also stopped the flow of experience. In trying to resist or ward off the undesirable or unacceptable parts of my experience, I would only succeed in fixing it in my field of attention, and indeed making it even more menacing. The judgment "I can't stand to see *that,*" kept *that* (whatever it was) right in front of me. When I ceased to judge it, that image merged back into the stream of visions. So my second lesson was: *stop judging* your experiences as good or bad—resistance fixes and magnifies the negative.

The next teaching occurred when my partner left the room, and I suddenly felt a strong desire to have her sitting next to me. Like a child wanting its mother I began to fret and get angry and was no longer aware of anything else, whether visions or bodily sensations. The feeling of craving and wanting something that wasn't there took over my entire attention and obliterated all other aspects of my experience that had been there. Here was incontrovertible confirmation of the second of the Buddha's Four Noble Truths: the source of suffering is craving. Since we only and always crave for something that we don't have at the moment, craving inevitably leads to constant dissatisfaction, the existence of which is the first of the Four Noble Truths. So my third lesson was: *minimize craving*, it tends to take over, removes you from your experience, and fixates your attention on what you lack. When I was able to release wanting the mother or the woman to be near me, I was swept joyously back into the flow of images and insights.

Thinking about the experience of the young child wanting its mother, got me thinking further about my mother, who, now in her eighties, was dying during this time. Thinking about her dying reminded me that I had been thinking about my own dying during past weeks, and this gave me the feeling of a bond of mortality between us. I said to my partner: "I've been thinking a lot about dying lately." When she asked why, I said, "I'm not choosing to think of dying, it just happens and I noticed it. I like thinking about me dying." When she asked me what I liked about it, I said, "it opens up my consciousness."

It was like the teaching about the effects of judging on experience: if I ward off certain thoughts or visions, typically those connected with death, decay, and destruction, they tend to remain lurking at the edges of my consciousness field. On the other hand, if I let in those so-called "negative" thoughts and images, I can accept life and death equally. Death is a normal and natural part of life, not its opposite. The Great Goddess, in whatever form, gives life through birth and takes life through death. During this whole experience, or since then, I had not the slightest feeling of wanting to die, or intention to die, or sense that I was going to die soon. Quite the contrary: in a later part of the journey, I had a vision of a six- or seven-year-old girl, wise with ancient wisdom, who was our daughter. This was a prophetic vision, as it turned out, since our daughter was born the following year. I was content with the knowledge that the time and manner of my dying, or the birthing of my child, was not up to me.

Accepting this, choosing to accept this lack of choice, made me feel a very deep peace, and a growing love—for the wife to be, the daughter to come, the three-fold Goddess, and the self.

In looking back at these experiences with ayahuasca, they represent initiations into ancient lineages and practices of healing and visioning. These healing visions, and the practices connected with them, were given to me, taught to me and shown, by the spirits of plants, of animals, and of the Earth herself. They have left me with an enduring, and growing, sense of profound gratitude and awe at the magnificent and mysterious beauty of Life.

We Are Experiencing the Joyful Phenomenon of Re-creation

CRISTINA SANTOS

In this account a twenty-nine-year old Shiatsu therapist and writer reflects back on her adolescent experiences with psychoactive drugs and later experiences with Buddhist insight meditation, comparing them with the profound union with trees and all life experienced under ayahuasca.

My work with entheogens began at the age of fifteen. My older brother and I and our best friend, Will, would take LSD and trek through the woods behind our house in suburban Connecticut until we arrived at the Pit: a bowl-shaped clearing in the trees beside a small pond. Summers were the best times for such adventures because we could stay out all night unshod and lightly dressed, spread out among the stiff, wild grasses of the Pit to gaze out on the stars. When I left for boarding school, I continued my exploration in the company of cannabis, PCP, and opium and began a love affair with MDMA (Ecstasy) that lasted well into my college years. In college in Minnesota, I added mescaline and psilocybin to my growing list of travel companions.

The summer break after my freshman year at college, Will told me of a Buddhist meditation course he had attended in Shelburne Falls, Massachusetts. He exhorted me to try it because "It was totally trippy!

You get to experience your body down to the smallest sub-atomic particle!" Of course, to one whose first love was entheogens, this promise of meditation bore wild appeal. That day I called the Vipassana Meditation Center and reserved my spot for the next ten-day course. What I found, after taking a ten-day vow of silence and meditating for twelve hours a day was not at all trippy. In fact, it was excruciatingly painful to sit still for one- to two-hour meditation periods. My mental agitation was extreme. I felt like a failure as a Buddhist, and though I had promised myself that I would see the ten-day course through to its conclusion, I swore that afterward I would never again meditate. On the last day of my course, in the nadir of my angst, I expressed my feelings to my instructor, Paul (a highly experienced Vipassana teacher and psychiatrist). His calm demeanor invited me to relax in his presence, and he suggested that we meditate together for awhile. I positioned myself in a half-lotus directly in front of him and began the technique of scanning the sensations in my body from head to foot. Moments into the meditation, I became aware of a firm, blue energy that was humming about my head. It arrived at my throat with a popping sensation, like uncorking a bottle of champagne, and I began to weep uncontrollably. I exited the meditation hall where over a hundred people were meditating in silence, and I walked outside into the chilled sunset air of the Berkshires. I was overcome by something that I now recognize as sorrow, a profound, ecstatic connection to the suffering of this planet. It was deep and rich, and powerfully satisfying to a hunger I had not been aware of until that moment. Buddhists call this *dukkha*, the direct experience of universal suffering that is the first step on the path to enlightenment.

Though I had sworn I would never again meditate, within days of completing that fateful course I embarked on a meditation practice that has filled over a decade with daily hour-long sits and annual ten-day courses. Vipassana, *dharma*, really, has been the greatest teacher of my twenty-nine-year lifetime. And at twenty, I gave up all drugs, at first because of my desire to explore Vipassana according to its pure guidelines that exclude intoxicants, and, eventually, because I felt that I no longer needed entheogens to explore or expand my consciousness. Last winter, after having achieved over a decade of pure dharma practice and

six months before my thirtieth birthday, the plant teachers unexpectedly sprouted up again in my path to knowledge.

We arrived at Coba at sunset, both fairly tired from our travels
across the Yucatan. My chemist friend, Burt, asked me if I wanted to try
the *pharmahuasca* [a combination of two synthetic drugs that mimics
the combination involved in the ayahuasca brew— Ed.] he had cooked
up in his lab. He explained that the trip would take about three hours,
then we could eat and go to bed. I assented, and we washed down the
clear capsule containing pharmahuasca (150 mg harmine + 100 mg
DMT freebase) with bottled water and relaxed in the cozy hotel room
awaiting her arrival. The trip began, as expected, in under an hour.
It was lovely: liquid colors with melting edges, deep ecstatic sighing
signaling the loosening of soma, and a sense that the roof had been
removed from my psyche availing infinite cool space above me. But it
was somewhat disappointing too. It felt like a drug. I felt intoxicated
as opposed to enlightened, inebriated instead of connected, and I was
bored and requested time checks long before the three hour mark. I
suggested that we go outside, feeling that if I were out by the lake and
under the stars, then perhaps I would have the experience of communion that I had sought by taking the pharmahuasca.

We walked to the end of a pier that extended into the inky waters
of the still lake. We lay down in silence just above the gentle ripples
and beneath a glittering minuet of stars. At the three hour point and
with nary a hint of communion, I said to my friend, "You know,
don't you, that there is a huge difference between taking this drug
and ingesting the plant material." It was a statement, supported by a
conviction, the origin of which I could not identify; but I was certain.
My staunch scientist friend replied with a knowing "Uh huh" that
said he had heard all the organophile arguments in favor of natural
substances and had no intention of defending his creation. At that
moment I spotted a tree near the edge of the lake that was illumi-
nated by a street lamp of the driveway of the hotel. I told him that I
wished to go over and see the tree. Our trip having ended and both
returned to the baseline, we walked down the pier under the indigo
Yucatan sky to visit the tree.

I approached the tree the way that I might approach a lover in a very intimate setting. I laid a gentle hand on her coarse bark and stroked her with affection, allowing the lines of my fingerprints to study each minute knot on her surface, caressing the smooth patches with my fingertips. Her gracious curves invited me to ascend her trunk. Once there, I leaned my weight into her bifurcating limbs and wrapped my arms around her largest branch. My heartbeat reverberated against her sturdy structure. Suddenly I began to weep. It was the ecstatic sorrow, dukkha, that I had experienced a decade earlier in the meditation hall. My chest heaved in tender spasms and tears rolled down my cheeks for an eternity as the suffering of the planet washed over me in its myriad forms. I felt a profound grief for those who lived lifetimes without the experience of true peace in their hearts and minds. As my mourning abated, I noticed that my fingers were lying over a knot on the branch that resembled a vulva. The branch bifurcated just at the edge of the ovular shape, resembling two legs, the whole image being that of a woman, upside-down, legs slightly parted, feet reaching to the sky. I began to weep once more, but this time for a very particular suffering: that of women who had been raped. For the past four years I had worked almost exclusively with survivors of sexual abuse in my private practice of Shiatsu. The tree was telling me that my work was evolving; it was no longer solely about accompanying women into their mourning and sorrow, but additionally about helping them to celebrate and to create. She illustrated this by drawing me out of my sorrow as I laid healing hands on the small tree vulva while directing my attention to another branch. The feeling of this branch was spontaneity and creativity. Joy. The images that paraded across my mind were of poetry, dance, sculpture, and laughter. It wasn't as familiar to me as the experience of dukkha, and I returned to dwell in that more familiar place of rich and heavy grief. Patiently and persistently, the tree redirected my attention to the joy branch, the world of creativity and renewal. Her message was clear: you know the dirge by heart. It's time to teach yourself a new song.

Gradually, I immersed myself in the rhythm of joy. I marveled at the tree, awestruck by this act of creation. I studied the intricate coiling patterns of her cortex, felt the rough wrinkles wrapped around smooth structure. I wondered at the sensual ascent of her limbs, stretching upward to embrace infinite space. I brought my lips close enough to brush against the rough patches of her bark and I whispered, "Look at how you've designed yourself. You're absolutely beautiful." Instantly, the lesson of creativity, in the broadest sense, was clear to me. We have all designed ourselves—these marveling eyes, these learning hands, this verdant planet breathing with green lungs—from pure energy. Some irresistible act of creativity willed us all into existence, and we are, from moment to moment, experiencing the joyful phenomenon of re-creation. Life, itself, is a masterful work of art and joy tickles the roots of this flowering creature.

My friend and I returned to our hotel room, he to drift off to sleep, and I to travel with the ayahuasca spirit for another eight hours. I experienced the tidal pull of the planet as I synchronized my exhalation and inhalation with Gaia's ebb and flow. I descended to a red-clay place in the center of the earth and below my navel, where Mother Earth herself spoke to me. Lovingly, she stroked worry from my brow and repeated the message of the tree: look how beautiful you've designed yourself to be. See how patient, gentle and kindhearted you've become in your twenty-nine years. She reminded me of how well I've loved, how deeply loyal I've been, and, most importantly, how I never once abandoned myself through the vicissitudes of my life. The message penetrated deeply, and I relaxed into the still acceptance of her unconditional love. For the first time in my life, I felt safe and completely joyful. I brought my lips close enough to brush against the soft cheek of my sleeping friend and I whispered, "Thank you so much for returning my magic to me."

Now the summer after my ayahuasca experience, I continue to interpret and understand her densely-packed eight hour lesson. I believe ayahuasca came into my path to reflect to me all the important work that I have done in the core of my soul—the moment to moment, assiduous attempts toward truth and beauty that no one but I can see. And I am certain that the plant spoke to me with such clarity and love because I have spent the last decade cultivating those very qualities in my meditation practice. Whereas my earlier drug experiences had been trippy in the extreme, now with the discipline and patience that grew out of assiduous, daily meditative work, the spirit of ayahuasca answered my call with the ripened voice of maturity. She came to remind me of my magic—the entheogens—and to encourage me to explore the synergistic relationship between my magic and my meditation on the dharma path and with my healing work.

A physician in his thirties, with moderate prior experience with psychedelics, recounts his initiation into ayahuasca, where he experienced the true meaning of "medicine" for the first time. Encountering both personal terror and visions of collective horror, he learned to pray again. Taking ayahuasca in the Amazonian rain forest, he marvels at the complexity and preciousness of life.

My first experience of taking the ayahuasca medicine was truly one of the most remarkable I have ever had. It brought me in touch with something very essential, something very deep within myself, allowing me to access a core identity that is normally obscured by ego and attachment. After the experience, I seem to have retained some capacity to enter into a calm and centered state of mind, even as the day-to-day storms continue to rage.

My expectations prior to taking the medicine were a cross between anticipating a profound healing and spiritual rejuvenation, and the fear that I would go stark, raving mad. In certain respects, both of these expectations appear to have been realized.

The hour or so before we began, I recall a time of mounting anxiety and excitement. Where I would decide to sit in the circle took on great importance. Although I questioned what I was getting myself into, there was no consideration of turning back. It was a great relief when we were finally ready to begin.

Walking over to take the medicine, I could still feel the conflict over what I was embarking upon churning away inside of me. It was therefore of some surprise to me that when it came to be my turn to drink, I did so very assertively, without any hesitation. I felt a sense of relief and a surge of confidence upon observing that the anticipated ambivalence was not there.

During the early stages of the experience I felt relaxed, drifting into a reverie of pleasant thoughts and images. My mental content for the most part remained in the here and now, securely anchored to my familiar ego identity. At times my imagery became eroticized, without any of the usual guilt accompaniment. At other times it became humorous, yet tinged with compassion. At one point I reflected that the experience seemed to be a rather easy one, somewhat like a mild MDMA trip. When the guide announced that about three quarters of an hour had elapsed, and that any one who felt they were "not quite there yet" could take a booster, my arm shot up, with a speed that was a bit surprising as I had earlier told myself that I was not going to "overdo" it.

Shortly after lying down from the booster I began to sense that I was now in for it. At first I began to feel some physical discomfort, initially gastrointestinal, followed by a generalized restlessness. I also began to feel some confusion, and a strange sense of detachment. As I became preoccupied with this rapidly increasing dissociation, I began to experience mounting anxiety and fear. The memory of a somewhat frightening mushroom experience a few months before crossed through my mind. I became involved in a desperate struggle to maintain some sense of control.

Lying there, trying to focus on my breathing, I reassured myself that the fear would pass, as do all things. Then I had the sense that my ego, my familiar identity, was beginning to fragment. I experienced a pulsating rush of ego attachments, weighing on me, heavier and heavier. Feeling that the weight of all my attachments was getting too heavy to bear any longer, I decided that this seemed to be a good time to throw up. I sat up, grabbed my bowl, and after a few dry heaves, began to vomit. As I was doing so, the thought crossed my mind that I was throwing up all my ego attachments, all of the things that were weighing me down, holding me back. Although I would not call it a hallucination, I imaged that the purging flow of vomit was transformed into a cascade of shimmering beads. After I was done, I lay back down, feeling considerably relieved. Although I was by no means done with fear and angst, this brought to a close the most terrifying aspect of the experience.

The experience continued to be strong and at times frightening. I remember at one point thinking of what Hermann Hesse had said in *Steppenwolf,* that the Magic Theatre was "not for everyone." The message I began to perceive was the need to let go of my attachments, to actually relinquish all that I identified with. I found it surprisingly easy to let go of some things, including my investment in being a doctor. Other attachments were harder to give up. One painful struggle in particular I can recall was that of letting go of being a "nice Jewish boy." For some reason this one was very difficult. I also recall, however, the very profound experience of completely letting go of life. I sorrowfully began to conceptualize that all of my attachments, all of my identities, were but transient phenomena and would inevitably end. I thereupon imaged my death in all its horror, including the rotting and decay of my body. I was able to very clearly see my skull crumbling into dust. Oddly, instead of the expected terror, this image left me with a sense of profound awe.

At this point, it is rather difficult to recall the precise sequence of events. Certain memories do stand out, however. I remember being flooded with BPM II imagery [associated with the second, oppressive phase of the birth process in Grof's model—Ed.] including being among masses of terrified concentration-camp prisoners, awaiting my turn for execution. I also imaged piles of bodies of South American peasants dying in ditches, after presumably having been executed by soldiers of a tyrannical government. These images left me struck by the profound suffering that humanity has had to endure. I felt tremendous sadness and grief, as well as rage that all this suffering had been caused by the acts of other men. Surprisingly, I also felt some compassion for the executors, in part because I seemed to be able, in a very disconcerting way, to identify with them as well.

As a rule, I don't pray. Yet, owing to the nature of the session, normal rules did not quite seem to be in effect. So I prayed and prayed, desperately. For myself, for all of us in the circle, for my family, for humanity. I prayed and I prayed. For deliverance from suffering, for healing. I began to cue into the chanting of the music, expressing my prayer deeply and sincerely. As I prayed, I began to feel uplifted, filled with a wonderful sense of calm. After having so recently experienced my death, it felt miraculous to be alive—truly alive—and in touch.

When the guide called for the circle to reform and for the talking stick to be passed, I had no idea what I would do. Yet as I took the stick, I felt a great sense of peace. Although as a rule I never sing (I've known since I was a child that I can't carry a tune), I sang. I sang my prayer to the spirit of the medicine that we all be healed. As I sang, I had the sensation that I had found my true voice. It seemed that this voice of mine was something I had lost or been forced to put aside long, long ago. To be able to access it again filled me with a great joy and peace.

I am still amazed that I had such a positive and profound experience, in spite of the terror and dissolution of self in the earlier stages. I also feel profoundly grateful that I had the opportunity to take the ayahuasca medicine. Its essence for me was its capacity to heal. The revelation that such a medicine does indeed exist is inspiring. I was filled with questions. How can this medicine be applied? Should it even be openly discussed or must it be kept a secret? As our world rushes toward an apocalyptic self-destruction, how could such a medicine as this be used as a vehicle for healthy and compassionate change? That the Amazonian rain forest now appears to be on the verge of extinction strikes me as no coincidence. What can be done?

As a physician, I commonly use and prescribe medication. Until this experience of ayahuasca, I had never experienced what a true medicine might be. It is a terrible shame that we are unable to share the secrets and powers of this medicine with the suffering people who come to us for help. I would like to believe, however, that a strategy could be implemented for the future that could facilitate such intervention. If our society is unable to incorporate such a change, however, it will be a sad world indeed.

I remain in awe of the beauty and power of the experience. I am the same person I was before the session, and yet I've changed. I seem to have taken away from the experience a sense of calm and connectedness that is very precious. I also feel that there is hope for myself and for the world.

Five years after my initial experience, I had the opportunity to take ayahuasca in a wild Amazonian rain forest setting. Far from the closest center of habitation, we made camp in a clearing on a high bluff, overlooking the river. As night fell, we were enveloped by the sounds of the forest, surrounding us on three sides. During our session, I had the experience, conveyed as revelation, that it was no accident or quirk of fate that we, as contemporary and sophisticated representatives of a world culture far removed from native peoples with traditions of plant-hallucinogen use, were encountering ayahuasca at this particular time in history. What I heard, or saw, or somehow grasped on a deep intuitive level, was that the spirits of the Earth were communicating to us through these extraordinary plants. Conveying that the collective Gaia-nature of this planet cannot much longer sustain its health and vitality in the face of escalating environmental destruction perpetrated by a world culture dominated by greed and aggression, the essence of this ayahuasca inspired communication was to wake up before it is too late and mobilize what forces are necessary to prevent the annihilation of nature and the obliteration of the life force it nurtures.

Later that night, while walking through the moonlit forest, we happened upon a tree with a leaf that, for all that could be determined, was in a state of perpetual motion. No breeze was blowing, no other leaf on a tree or adjacent bush was moving. And yet, this particular leaf, without evident cause, continued to gyrate, rotate, and turn continuously, long into the night. Though antithetical to the reductionist mindset, we could only but confirm that nature was speaking to us. Indeed, nature was alive!

Traveling further up river into even more remote regions, we were taken by canoe to a small island. There were six of us, including our local guide. With the sun setting to our backs, we faced the river, listening on a boom box to a tape of ayahuasca chants of the Kulina Indians.

About one hour into the experience, with visions of subtle yet immense beauty beginning to crest over us, the tape came to an abrupt end and clicked off. Silence, as we faced the wide river. The sky, resplendent with the setting sun, turning to dusk. Sitting on the narrow beach, I slowly turned my head to the left. At the end of our group sat my friend J. And yet, beyond, and equidistant from J. as we all were from each other, reposed a seventh and newly arrived presence: a dark and still vulture, also sitting quietly facing the river. As my eyes remained riveted on this aviary apparition, unsure of its corporeal nature, J. turned to me and perceived my attention to be directed somewhere to his left. Slowly he swiveled his head to align with my gaze, and then even more slowly turned his head back to look at me, eyes wide open, mouth slightly agape. Later into the night, long after our session had ended, J. and I spoke about what we had seen. Without such corroboration from my friend, I might have understood my perception to have been a hallucination, somehow conjured up from the murky depths of revealed primordial unconscious. But J. had also seen the bird, sitting to the far left of our row of enraptured sitters, joining us in our communion with the spirits of ayahuasca, the vine of the dead.

Since my first experience with ayahuasca, I have had the opportunity to participate in a fairly large number of ayahuasca sessions in North America, South America, and Europe. Settings have varied, from tightly structured South American syncretic churches, to contemporary North American groups utilizing shamanistic models, to more free-form experiences with small groups of close friends. My personal encounters with ayahuasca have ranged from accessing heights of astounding transcendent bliss, to plummeting into the fragmenting depths of nihilistic despair. I have traveled to other realms and seen visions of extraordinary beauty and complexity. Knowledge and information, contained in the core of the experience, has swept through me. I have been catapulted to a domain of being other than my self, more akin to the True Self. I have stood humbled in the face of its immense otherworldly power and have dissolved in the embrace of life-affirming ecstasy. These encounters have provided a learning experience of extraordinary depth and profundity. Ayahuasca is a teacher that guides, reveals, and makes manifest the ancient wisdom carried in our souls. At its heart is a moral compass that unerringly orients and directs down a path of simplicity and truth. I have learned that with ayahuasca as a teacher, I may not always get what I want, or expect, but I invariably experience what I need. Painful or not, this work with ayahuasca has been a blessing, whose power to heal and enlighten continues to facilitate positive change long after the acute effects of the chemical compounds have departed.

Breaking from the Bondage of the Mind

Kate S.

A woman artist in her forties recounts an experience with ayua-huasca in which she was confronted with a vision of the inescapable round of sex and death ("fucking and dying"), in which we humans as well as all other life are involved. As she worked through her initial resistance to this seemingly primitive inescapable force, she attained to a liberating vision of light emerging from the death of form.

I began early on a Saturday morning. After downing about ¼ cup of the tea I laid down and listened to music. It was very difficult for me to drink the tea. I felt nauseous immediately. My effort was to keep the tea in my stomach as long as possible. This I did for about twenty minutes. By this time I was hallucinating heavily. My sitter helped me to the bathroom. I was barely able to stand. I went to the bathroom and proceeded to vomit. I felt an incredible sickness throughout my whole body. I was finally able to return to a lying position back in the room and continue to listen to music. The music I was listening to was from India. It greatly influenced my hallucinations.

As I lay there, images flooded my mind. I began to see many statues. They looked as if they were made of some sort of sandstone. They were a sandy orange color. They were about six to nine feet tall. Most of the statues had faces and bodies and they were all formed in such a way that they were all connected, one to the next. I turned my gaze to look at them and snakes began to come from all the opened orifices of the statues. The snakes were black. There was a very continuous circular movement of snakes. If I would open my eyes and then close them again the movement of the snakes would stop and then in a couple of seconds their movement would begin again. Soon the forms of the statues became more ornate and very Indian-like in their design. I had the distinct impression that the art style I was seeing was because of the music I was listening to. I felt that the energy of the Indian music was forming the type of art I was seeing.

I had the thought that the reason certain cultural or ethnic art forms appear is because of the planetary energy in the location of the origin of that form, and that the music and the art were intricately connected and reflective of the energy of the planetary location of their origin and the energies that exist there. The statues became separated from each other. They were covered with jewels that sparkled. Especially small red and blue crystals. Still the snakes came, but by this time they were also beginning to be more ornate. They had beautiful diamond-shaped patterns over their skins. They stuck their tongues out, and I was aware of their bodies' exquisite grace and the flirtatiousness of their tongues, inviting me nearer into their energy. Around this time I asked my sitter to stop the music. When this happened the appearance of the objects before me suddenly shifted.

Soon the statues became human and animal figures that were alive but still had a sort of statuesque appearance. I became aware of dense jungle flora around the figures. It was very green and very alive with the movement of life-forms. Soon the figures became involved in a sort of giant sexual orgy. In my mind I heard myself say "It is all about fucking and dying." I watched two humanlike creatures having sex, and I became aware of the rhythm of the sexual act. Suddenly there were hundreds of entities fucking. It seemed that the sexual act had no attachment to any type of passion, but rather had its own will and was based entirely upon animal instinct. It was to be and would exist regardless of anything else. I understood that the sexual act was guaranteeing the continuation of life forms on the planet and also generating a kind of energy that was being given off by the entities involved. These were its purposes and its only purposes. Pleasure or personal relationship was an insignificant side effect. It was not a thought-out sort of reality, but a law of the planet. There was a certain feeling of violence connected to the unrelenting need for the sexual act to take place. I continued to be barraged by fucking life-forms.

I heard the sentence, "We are less than we think we are." I became aware of an inner fear that the spiritual light that we are drawn toward is a sort of ruse only there to make our unimportant position of the planet more bearable, and that somehow our energies are being harvested by a greater power or force that needs the kind of energies we generate not only through the sexual act, but through the energy of violence, and that we are oblivious to all of this.

At this point I became aware of a large skull and chest in the distance. It was rather like a statuesque bust of a person but all of bone, no skin. It had a sort of luminescent, pale blue tone to it. I approached it and stood watching it. Suddenly the top of the skull exploded into a million little shards and light streamed out from it. I was struck by the beauty of the sight. It was a spirit connection. I thought, "Yes, this has really blown my mind and my way of thinking about who I am and who we all are." After this point my hallucinations began to taper off. I again became aware of my present existence.

Reflections Two Years Later

My immediate sense of this experience was not especially positive. I felt confused about the meaning and relevance of many of the things I had seen. I had a feeling of betrayal in terms of the spirit realm and doubted its authenticity. Later, when I had time to integrate my experience more, I realized that I have had a long distrust of the Spirit, and that the suspicion of the Spirit being a false thing was an old familiar theme dating back to not trusting the security and connection of my family or of a God that would not protect me from the pain of my early childhood.

I had a sort of claustrophobic reaction to the idea of existing in the realm of "fucking and dying," and yet I understand now that on one level that realm really does exist. When I was in the experience I found myself not wanting to see the meaning of what I was viewing. It seemed too hard to bear. I did not want to merely be here to fuck and die. I did not want to be so ignorant of the reality of my existence that I was, as Gurdjieff would say, just here to feed the moon. So for awhile I had the feeling of being stuck in a not very appealing place. This feeling was a reflection of my own internal state at that time. I do feel that this experience made a rather dramatic shift in me in relation to how I look at human purpose on the earth. I now hold the underlying belief that if one's life is not lived somewhat consciously that the sum total of it may just be the "fucking and dying" that I saw so clearly in this experience. And that "fucking and dying" is fine, but I want more in my life.

I have spent much time in thought about human existence, about what the spirit is, and why our spirit would chose to indwell this human form on this planet. We are very complicated beings, and the big joke is that we know so little about ourselves even though we think we know a lot.

The exploding skull was perhaps the most interesting part of this experience to me. I did eventually draw a picture of it. Initially, looking at the skull captured me because of its stark beauty and also because it seemed to emit a very healing, blue light-energy. When the skull exploded it gave me a surprise and a great sense of freedom and joy, a sort of breaking from the bondage of the mind and of the human form. The light that emanated from it after it shattered was the kind of white light that one sees in spiritual experiences. I have seen this light before. It has a brilliance that seems to pass through the viewer. It seems to touch the very essence of one's Self. One recognizes it. I believe that it is the light of the Spirit.

Now and again I think about this experience. It seems to me that I traveled down the road of one of the most pressing questions I often ask: What is the nature and purpose of our existence on this planet? In the end, the body was merely bones and what was left was the life of the light.

A Vision of Sekhmet

GANESHA

An actor and body therapist in his forties relates his first experience with ayahuasca, in which he unexpectedly encounters a vision of the ancient Egyptian goddess Sekhmet, of which he had zero prior knowledge. Along with this vision came detailed memories of having been a priest of this goddess and the nature of her teaching. Subsequently there also occurred visions of Tibetan Buddhist deities. The connection with the spiritual healing power of Sekhmet deepened and strengthened his therapeutic work.

I was in the High Desert of Southern California for my first session with the Circle and what would be my first ayahuasca journey. The Circle was a group of spiritual journeyers who used hallucinogenic plants in a sacred way for vision and healing experiences. My intention for the upcoming medicine journey was to seek a vision that would energize and fine-tune my spiritual path, and enable me to walk in balance on the Earth in a way that would benefit all my relationships. I also sought to have knowledge already learned to be transmuted into understanding at a deeper level.

About a dozen strong, including a guide and a sitter, we had arrived at our small desert house on a Friday night and had spent the following day in the surrounding desert on solo vision quests, shedding our busy lives, communing with nature, and fine-tuning our intentions. At the end of the day, we gathered once again in the house and began to prepare for the journey. The circle ritual was opened with an invocation of the four directions and an honoring of the spirits of the place. At about 7:00 PM, I was given 90 ml of a liquid preparation of ayahuasca. It was not unpleasant tasting, a rich chocolaty molasses flavor. At first, I experienced the medicine only somatically. I felt nausea, but it was faint at first. I breathed into it as the guide had suggested and "rode it." I found myself using my hands to channel light to my body and my chakras. I felt a connection to the earth and a sensuality, similar to my first MDMA experience with a therapist the previous month. I wasn't sure what to look for in terms of a vision. The guide had said it might not be visual, but a combination of thought and awareness patterns. The first hour was mostly about not giving in to the nausea and keeping the medicine down so that it would be effective.

Around 8:30 PM, I was given a 10 ml booster and almost immediately threw up. When I came back to my place in the Circle, I focused my awareness more deeply into my brain to feel the effects of the medicine. I sank deep within the actual substance of the brain, experiencing there a serpentine consciousness in the physical form and experiencing electrical currents running through the brain in the form of little serpents. The close physical proximity of these serpents to the *ajna* (third eye) and crown chakras made me think how curious was this juxtaposition of pure consciousness and serpentine substance. I observed that these two opposites were close, but not interrelating at this point.

I started having sporadic light visuals, but nothing sustained. I still didn't know how I would perceive my vision or if I would even have one. Then I caught a glimpse of what I thought was a large eaglelike bird flying high in the sky of my consciousness. I somehow knew that his name was Garuda.

It was around this time that the first round, or speaking session, was called by the guide. Smoke was passed in a pipe, an herbal combination of cannabis, datura, damiana, and dried amanita mushroom, and with one draw, I felt the ayahuasca really kick in. Marijuana had always been a good medicine for me, an ally that would put me in touch with some degree of shamanic perception. It was after this smoke that I began to know the nature of ayahuasca.

I began to become aware of the presence of a Being. The name *Sekhmet* came into my awareness, almost as though it were being whispered to me. Through this faint impression, I was only aware that Sekhmet was female and that she was Egyptian. I went on to other experiences, but the question "Who is Sekhmet?" was burning in my consciousness. Several times the question was so loud that I almost asked the guide who she was, feeling that he would surely know the answer. I never did ask him, but the question in my mind went out with a power that had to call in an answer. I became aware that she was a goddess and thought I must have seen her in a book I had been reading a few days earlier that had contained a plate depicting Egyptian deities with their names printed underneath their images. I could see her on this page as a small hieroglyphic figure of a cat-headed woman, standing, facing sideways. Very clearly printed under her depiction was the spelling of her name: Sekhmet. The name was familiar. It was as though I somehow knew from the distant past who she was, but had forgotten. I began to see images of this goddess carved in black stone, seated on a throne. There were hundreds of these statues floating around in my mind's eye.

I was also aware of the presence of other goddesses. Far below me, or so it seemed, I saw the dark figure of Kali. The visual of her was slanted, as though on a different plane. I didn't approach her.

All of a sudden, the full presence of Sekhmet flashed into my perception. She was revealing herself to me in all her glory. Just as suddenly, I had to vomit again. The purge was lengthy and full and purifying. When I rejoined the Circle, I tuned back into the vision I had been given of her. A tension had been released with the purging, and now I could lie back and let the details permeate my awareness. It was as though I had downloaded the vision when she appeared to me, and now I could access the information.

In the vision, she was seated on a throne in a cavelike temple within a desert mountain. The cave was illumined with golden light, and it seemed that the Sun itself was behind her throne. Her face was that of a lioness, and she exuded great beauty, warmth, and beneficence. As I approached her, I experienced other beings, her guardians and attendants, beckoning to me, drawing me forward, and I felt the mountain opening up to receive me. When I reached her throne, she was surrounded by these guardians. Everything in the temple was infused with golden light. I was told telepathically that I was a priest of Sekhmet.

The next thing I remember is being given an initiation, a death rite passage. Sekhmet, it seemed to me, had a place of authority in the realm of death, an honored place in the underworld. I began to sense that she was present at the moment of death. And I knew that I was about to die. I checked in with my inner knowing to see if this was going to be ritualistic or actual death, and I was given the green light to give in to the experience.

I became Osiris in his tomb. Sekhmet was present. She had brought Death. She was Death's attendant but more than a mere attendant, for Death seemed to be in her presence as though she were its ruler, its mistress. I was lying on a death couch. I began to experience dying. I knew that it was my ego dying and that my mind was emptying out, stilling. Soon there were no thoughts left, only consciousness. My body was now on a barge floating down a river. It felt like the Nile, but it was known as the River of Letting Go. My body was dead, but I remained in it. My hands were folded over my heart, my mind was still, alert, listening. Above me I felt the presence of Horus as a hawk. I knew that I had the potential to rise up and become him, but instead I heard music, with the name Ishtar being softly sung. Ishtar seemed to be synonymous with Isis, and she had come to bring rebirth. I felt life starting to course through my body. It came from Isis-Ishtar, but it also came from Sekhmet who had been present throughout the experience. Indeed Sekhmet was running the whole show. Just as she had held the power of death, she was now the giver of life. Her mystery was that she represented the Great Mystery, which I was now experiencing as a mystical nonboundary between life and death, the simultaneous coexistence of both the presence of death and the force of life. And now she was making my body into a lion. As life poured in, so did the qualities of a lion. In fact, I felt the force of life as a lion. They were the same.

Once again I received the telepathic message "Priest of Sekhmet." I didn't know if this meant I had been her priest or that I was to become her priest, but the sense was that I had been and would always be. I was given a teaching to impart to the Circle, which I did during the next round. The gist of it was this: one must bring death into life. Death must be embraced, for life follows death and only if one totally lets go into death, can one fully live. This is true at the end of a lifetime, at the end of any cycle, at the end of each instant. I was told that if death were faced within oneself, there would be no killing, no wars. That is only acting out or projecting onto others the death that is denied, rather than embraced, within oneself. As she taught me, I felt her entering my body, and I opened to receive her, to become the Lion Goddess. I felt trans-fused by her spirit. My face became her face, the face of the Lioness.

At this point in the evening, the guide put on the music of the Gyoto Monks of Tibet, and I felt compelled to sit up and face East. I went into full lotus quite effortlessly and felt strong tangible waves of energy com-ing into me as teachings of the East. It felt nurturing and completing since I had been more recently focused in my life on teachings of the South, West, and North. The drums and the deep sounds of the music kept taking me deep into the Earth, deep into my root chakra. I felt the surge of kundalini and felt the Tibetan music was designed to awaken the kundalini power and to balance all the centers so that they would open to this power. Eventually I felt the serpent totally inhabit my body. I was the Serpent Goddess, Kundalini herself, my head a serpent's head, rotating to look around, my tongue darting out of my mouth like a snake's tongue.

Throughout this part of the session, I felt the balancing of male and female energies, the dance of consciousness and substance. It was so easy to leave the Mother and be drawn upward toward the sky, but the drum was like the heartbeat of the Mother, serving as a reminder to stay connected. I felt the grieving of the Mother when I would be away from her, and felt my own mother's grief at our separation. I experi-enced myself from the perspective of pure masculine spirit and my love of form. I jotted something down in my journal and later found the words, "Yes. I came from the stars to love you, Mother of All Forms, in the body of this Earth, Gaia."

At one point, the guide led the Circle in invoking spirits. The doors to the outside blew open and I saw spirits entering on the wind in waves of rainbow light. They assembled around the circle of people.

Toward the end of the session, I became aware of an energy parasite in my splenic plexus. I unified with it and became its consciousness, my face distorting to look like it, and, using my breath and an extremely strong will and concentration, I reached my etheric hand all the way into my physical body, took hold of it, and pulled it out. It was in the form of a scorpion. I dissipated it into the air around me. I felt some residue still within my body, and since I wanted to see if I could do it again, I repeated the operation. At that point, the guide and the sitter started to drum. The sitter was near my feet, and the guide was in his position behind me and to the right. I drummed with them, using the skin over the etheric wound as a drum. The drumming aided in the healing, and this area of my body proved to be quite resonant.

The full effects of the ayahuasca lasted about five or six hours through the closing of the circle and then lingered on somewhat for a couple more hours as the group broke their fast with hot vegetable soup and began to share experiences.

When I returned home from the Circle, I immediately went to the book I had been looking at some days before, to find the picture of Sekhmet. I found the page, but no depiction of Sekhmet. I did find the god Set and the goddess Nepthys and surmised that in my hallucinogenic state I had synthesized their names to create the name "Sekhmet." I checked all my other books that could possibly contain some reference to her and found nothing. The next day, as I sat down to write up a record of my journey, I was starting to feel that this Sekhmet was nothing more than a product of my ayahuasca-induced imagination. I stayed true to my vision, however, and wrote up the experience.

Later in the day, I went to the Bodhi Tree Bookstore to scour their Egyptian section in hopes of finding some reference to this lion-faced goddess. After searching for about an hour to no avail, I came upon a book called *Her Bak*. I could find no reference to her in the index or in the text, but as I was about to place the book back on the shelf, it fell open to a plate of about twenty-five gods and goddesses. No Sekhmet. But I turned the page to another plate and there in the midst of another twenty-five or so deities, looking somewhat insignificant, but definitely in existence, was Sekhmet, a woman with the face of a lion. I was elated. I had found her.

During the course of my research over the next few days, I was to discover that Sekhmet was one of the most ancient of deities, known as Lady of the Place of the Beginning of Time and also as One Who Was Before the Gods Were. Although having a ferocious wrathful aspect, she was also a renowned healer because of her knowledge of magic and sorcery and is said to have introduced the art of medicine to Egypt. Many Egyptians felt that they were created by Sekhmet and Horus, and that Sekhmet was their protectress and spokeswoman in the underworld. Rameses II believed her to be the mother of his soul and, interestingly enough, had hundreds of large black statues carved in her image. She is always depicted with the sun disk crowning her head and usually with the uraeus, or cobra, around it; in her hand she holds the ankh, symbol of life. In one book, I found a reference to the priests of Sekhmet and discovered that for centuries they were regarded as the most potent healers and magicians of the ancient world, very often performing their healing from a trance state.

As I read about Sekhmet and assimilated my experience with her, the understanding that formed in my consciousness was that Sekhmet is a Great Mother Goddess, one that spans all time. With the sun disk at her head and the snake around it, she symbolizes the serpent power of the root chakra having risen to the crown. Thus, she encompasses both heaven and Earth and demonstrates the way to unite the heaven and Earth of our own nature, Spirit and Form, through the awakening of the kundalini power in the *muladhara* chakra and its arising to the *sahasrara* chakra.

I found references to lions and serpents in various mythologies other than the Egyptian, the most potent being in regard to Senge Dolma, the lion-faced *dakini* who transmitted the tantric teachings to Padmasambhava, the founder of Tibetan Buddhism.

In the days following the journey, I felt the presence of Sekhmet quite strongly. Exactly one week after the initial experience, I once again felt her infusing me with her spirit. I seemed to become a lion, and my heart chakra was expanded and filled with the power of the Sun, as I heard the words "Lion Heart." Two days later I was given a teaching in a dream wherein the Lioness Goddess had three attributes: life-giving, nurturing, and destroying, similar to the Hindu godhead of Brahma-Vishnu-Shiva.

This journey proved pivotal in my life, just as I had intended it to be. As I asked for, I was given a vision for my path. Through my healing experience with the etheric parasite during the circle ceremony, and through my discovery that priests of Sekhmet were healers and magicians, I embarked on a path of shamanic healing. The day following the journey, I met a woman healer in the desert who was to become my lover and teacher of healing. Together we would invoke the presence of Sekhmet in our healing work. I would very often feel Sekhmet within me, receiving her transmission and "becoming" her in learn the way of her healing. I had done healing work before encountering Sekhmet, but she took me deeper. I found myself going into a trance-like state during healing sessions, which provided intuition in "reading" the patient and brought a certain touch of magic to the healing. Visions would paint a picture of the area of the body that needed attention and lead me through pathways of inner landscapes to remove obstructions, channel light, and restore health and balance.

Besides her nurturing and healing side, the warrior side of Sekhmet has been equally valuable to me. Since my vision, I have felt a strong sense of being under Sekhmet's protection. Her wrathful aspect in mythology I feel as the ferocity of the mother lion caring for her cubs. I was later to receive the transmission of Senge Dolma from a Tibetan *rinpoche*. The mantra is all about protection. I use this harm-dispelling mantra whenever I feel threatened.

My research of Senge Dolma and Padmasambhava led me to explore Tibetan Buddhism to see what secrets the lion-faced dakini might have revealed to the man who brought Buddhism to Tibet. I later came to understand the flight of Garuda, which I witnessed above my head in the journey, to be a metaphor for one of the highest teachings in Tibetan Buddhism, that of dzog chen, the Great Perfection, the view of nonduality. While visiting Nepal, I felt the powerful presence of Senge Dolma, and the day before I left a Tibetan monk who had befriended me gave me an delicately painted scroll of the lion-faced dakini. And Kali, the dark goddess who I glimpsed below me during the journey, was to visit me again when I returned from Nepal, on the wings of a different plant teacher.

One summer solstice, five years after my initial vision of Sekhmet, I traveled to Mount Shasta in northern California. After doing ceremony with two friends on the mountain, I looked up to see a large, substantial cloud moving toward the Sun. The cloud formed into a face with the Sun at its crown chakra. It was the face of Sekhmet as clear as though it had been painted. A woman's body formed beneath the head. The eyes were orbs of sunlight, openings in the cloud, and the countenance of the smiling face was of love and benevolence. Although the wind was blowing, this face did not change for about five minutes and was witnessed by the two friends with whom I had done ceremony.

My subsequent journeys with ayahuasca, although beneficial in terms of both healing and vision, were never as powerful as that first one. This journey gave me a major life vision. It came at a crucial turning point in my life as I embarked on a new path of shamanistic spirituality. The setting of this phase of my life combined with the set of my strong intention for new direction to bring about this phenomenon of encountering a goddess who would inspire, teach, and escort me into new realms of experience.

It is of interest that after ingesting a South American jungle hallucinogen in the middle of the California high desert, my vision would be primarily from ancient Egypt, with Hindu and Tibetan Buddhist overtones. The spirit of this medicinal plant transcends cultural, religious, and geographic boundaries, transporting one into the collective consciousness realm of all spiritual metaphors, into a limitless sea of cosmic possibility.

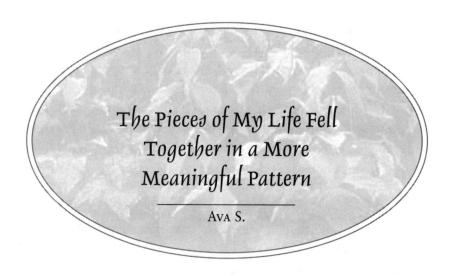

The Pieces of My Life Fell Together in a More Meaningful Pattern

Ava S.

In this account, a psychologist and dance therapist in her fifties describes how the ayahuasca helped prepare her psychically for uterine surgery, as well as giving her some deep insights into self-limiting patterns in her relationships. The format of this group ritual involved a four-hour solo vision-quest walk in wilderness, prior to the circle ritual with ayahuasca, which took place inside.

The following is an account of my third trip with ayahuasca. The first had been extremely visceral, and the second milder. I felt that I was in the midst of an ongoing dialogue with the ayahuasca spirits.

We began with a meeting on Friday night to declare our intentions. Mine was to heal; I was awaiting uterine surgery and wanted to work with the meaning of my symptoms. Since ayahuasca was such a physical experience and since I had already felt a strong relationship with it, my questions were: "What is at the root of my physical symptoms, and what is necessary to bring about healing?" Clarity and commitment of intention were necessary to give the ritual power. Our guide had reminded me to make my own healing a priority.

On Saturday morning we began the fasting vision-quest walk. We traveled out to a wilderness area in the desert. We were asked to focus on our intentions, on the qualities of the four directions, and to let the nature spirits of the desert guide us. I found myself drawn to the southeast, in the direction of open space and wide horizon. I happily followed an arroyo for about an hour, with little desire to climb into the rocks or sheltered spaces. I then found a rock ledge that offered a panoramic view of the valley, while still providing partial shelter. On this rock, I took off my clothes and basked in the warmth of sun on skin, and skin on warm rock. I felt close to the lizards and remembered how I used to keep and raise lizards when I was young.

I was also fascinated by textures: dead joshua trees with charcoal lattice-work bark, black and brown dappled rocks, subtleties of brown, black, and silver. I saw images that have been repeating themselves to me: swirling tunnels of dark browns, with yellow-eyed wolf faces emerging, fading into owl, boar, and bat faces, all textured with feather, whisker, and animal hair. The desert seemed to be preparing me for the ayahuasca, the open spaces and heat put me in a body space, while the vivid colors and textures presaged visions I would continue to have of fur, warmth, and texture, like the dappled brown and silver texture of this warm rock. Brown and silver, a new color combination for me. Melting into warm rock, naked skin on rock, dozing like a lizard. Warmth, texture, my being as a sensation of nature, ageless; no thoughts, just spaciousness and peace.

After we returned to the house in the late afternoon, we began the medicine circle ritual. Invoking the guidance of the spirits of the four directions, the personal and tribal ancestors, and the animal, plant, and insect world, set a strong circle and safe space for the group. We each then put a special object onto the altar in the middle and called on the spirits for help in following our intentions.

At first, the ayahuasca was visual for me. I saw the swirling browns and blacks, the animal faces and lower world imagery, the depths of browns and earth colors. When the music changed, I saw other colors—turquoise and pink—and playful shapes. Then a requiem came on and I found myself sobbing, feeling the weight of universal grief, of grief for the planet and for recent deaths and injuries in my family. After some time, the imagery faded, and I was mostly feeling.

Then the nausea came. I felt long and sustained waves of nausea, unlike the usual sharp and quick pangs of previous trips. I asked myself "What makes me nauseous?" and saw scenes from my life. My own life is making me nauseous; something is not sustainable or wholesome. I'm divided; something needs to come clean. As I bent over the pots to vomit, I felt my intense struggle to hold on, not to let go. Images of relationships and parts of my life that must be let go came to me; they are draining me just as my fibroids are an energy drain.

These images concerned several men and relationships I maintain, and especially images from my own past. I felt how in love with my father I still am, and how he is such a difficult act to follow. I know I needed to let go but fought the nausea and the imperative to let go as hard as I could. Each time I rationalized and thought about how I might avoid letting go of a particular relationship or connection, a new wave of nausea would come over me. I experienced a struggle between old mind patterns and a very eloquent body. Even though this was (and is) really a life and death issue, I could feel how part of me wanted to make light of or avoid the definiteness of letting go. Although I'd been told this many times, I *felt* for the very first time how much I was still in love with my father. I also saw how the secrecy of my relationships was nauseating me and tried to see how I could break the code of secrecy.

The next morning I felt very tired but more peace, less struggle; just naming what for me is a huge "family secret" was somehow liberating. Naming this complex in such a straightforward way in front of my peers helped me to get over a certain shyness and helped me see how otherwise unexplainable or irrational behavior made sense. Much as I had always resisted the reductionism of pointing to one variable as the determining factor of a life, I nevertheless felt a definite clarity and relief when I could name this one. And despite years of psychotherapy, analysis, and even being a psychotherapist myself, the picture had never been so clear, or so viscerally felt. As pieces of my life fell together in a more meaningful pattern, I felt a sense of peace. Being in love with my father was not a sin, and I did not need to feel guilt over it. What I did need to do, however, was to clean up my actions, so that they would not be unwholesome or unsustainable.

A significant part of the ritual was integration—bringing the insights and images into real life. At the closing, we were asked to reflect on the following questions: What was your intention? How did your experiences in the desert and night connect to this question? What will you take back to the world?

My intention was to heal the fibroids and prepare for surgery, to get to the secret of the bleeding and coagulation. My experiences in the desert and the night gave me images of death and life, enabled me to grieve for the several recent deaths, and gave me a strong physically-felt image of purging and letting go—letting go of my father and other relationships in my life that are like my relationship with him. What I took back to the world was an intention not to keep acting out this particularly unhealthy pattern, and to take even more responsibility for my own health.

Although I wrestled with the problem of how I would concretely implement these intentions, I also had faith that the ayahuasca spirits would help me. Since the learning was so in my body, and since I so trust bodily ways of knowing, I knew that any changes in my life would reveal themselves as I went along.

Sure enough, I found myself easily doing healing rituals to help prepare for surgery. With little resistance, I stopped drinking wine, prepared my teas and herbs, strengthened my body, and took pleasure from taking the time for these rituals. More important, the people with whom I needed to change relationships just happened to call, and distance themselves for other reasons, like not being able to travel. I didn't have to let go, or to make a drama of letting go, but life just took care of itself. Further, my father called the next day. Without rehearsal, I managed to tell him simply that I'd always loved him, and he was able to return the feeling. Although there is much more that could be said between us, I felt that this was enough, and could let him go in peace.

As I got ready for surgery, I was aware of how I used to be afraid of anesthesia and of losing consciousness in a hospital. I realized too that I hadn't had surgery since I was very young. Remembering how naturally I rode the changes brought about by the ayahuasca and emerged stronger, I was no longer so afraid of losing consciousness. Preparing myself with relaxation tapes and the help of loving friends,

I went through surgery smoothly, with almost no bleeding and a very fast recovery. The surgeon was amazed at my recovery process; I know that it had to do with all the psychic and physical preparation I went through. Within three days the stitches were out, and I am now recovering strength quickly.

I feel more calm and confident about my life, about keeping it on a healing keel, and can feel much more distinctly those people and projects that sap energy, as against those who give energy. I have been careful to keep healing thoughts and images in the foreground and feel the results. And in my movement-therapy work with women with cancer, I feel that I have more healing energy and can see them respond. Without having to share details of this healing journey with them, I am able to bring them some of its energy and intention, and together we heal. I know that ayahuasca, with its strong physical presence, was a big part of that healing for me.

Reflecting back on this experience several years later—many changes continue to happen on a nonverbal level. Unhealthy relationships have been purged, and I feel younger, fresher, lighter. I am able to take more nourishment from life itself and avoid being drained of energy. My professional healing work has grown, and I am involved in several exciting new projects.

On the whole, I value how the ayahuasca experience changed my life energetically, nonverbally, and enabled me to make new and healthier choices.

Knowledge Was Graciously Invoked in Me by the Plant Teacher

OREGON T.

In this unusual experience, a college philosophy professor in his forties obtains confirmation of intuitions of a prior lifetime as a student of an alchemist magician in Elizabethan England. Seeing the parallels between that life and his present life, he rediscovers certain mathematical relationships and also obtains liberating insight into personality patterns.

My intention on this first experience of ayahuasca was remembrance: during the past several years, I have received fleeting intimations and vague recollections of a life in sixteenth-century Elizabethan England. Much of my study, writing, intellectual pursuits, including even my Ph.D. dissertation topic, have been colored by the remembrance of that life. Following a series of out-of-body experiences, combined with specific details supplied by a gifted reader of the akashic records [according to occult philosophy, the collective astral and mental memory banks—Ed.] I was led to the recognition of the identity of this individual—Robert Dudley, the Earl of Leicester. Recognizing the uncannily detailed resemblances in the proclivities and interests of my present ego-personality, I wanted to seek communication with this previous life.

The intention was not to simply satisfy the ego by stretching consciousness into the past. Rather the concern was to get a handle on the soul's purpose in this incarnation, by remembering both the joys and frustrations of a prior lifetime. Specifically, I wanted to see if I could recapitulate the esoteric mathematical knowledge garnered in that life; if I succeeded I wouldn't have to reinvent the wheel in this regard. To anticipate what the spirit of ayahuasca showed me was consistent with my intention, but at the same time surprising in the lessons learned.

My initiation began with the proper fasting and a preparatory half-day solo vision-quest walk in the mountains, communing with the spirits of nature. We each consumed a very small amount of San Pedro cactus that served to heighten alertness and one-pointedness on our vision walk. I climbed high into the mountainous terrain traveling in a westerly direction. I noticed that birds were appearing to me in pairs; I interpreted this message to point me to the importance of love. For some time I watched carefully the dancing and jumping movements of a small fly. I was hoping for the appearance of a significant power animal, as practiced in shamanic traditions, and was beginning to get somewhat frustrated. I found a secluded spot amongst some pine trees and laid down to observe the birds some more. Finally I got to the point that I thought nothing was going to happen.

Just as I was about to give up, a lizard leaped out from under the brush right between my legs and onto a nearby large rock. Our eyes locked. As we gazed at each other, I was not sure who was mesmerizing whom. You might say that it was mutually hypnotic. The communication was telepathic, consciousness to consciousness. However, I did say some soothing words. Eventually I slowly approached him and was able to touch his back gently, petting him. I had taken my camera and recorded my finger stroking his head and back. He also allowed me to take several close-up pictures. Before long it was time to return to the group, so we parted ways.

Later that afternoon we drank the ayahuasca tea. I did not find the liquid preparation disagreeable at all. In fact, I took it with great relish. We gathered together in the vision circle, around an altar that included sacred ritual objects that each of us had brought along. There were fourteen of us along with our guide and his assistant. With my bowl for the subsequent purging next to me, I lay down placing ear plugs in my ears and a mask over my eyes. Before too long, I began to see within my mind's eye undulating, circling, spiraling formations in a panorama of kaleidoscopic colors.

I soon had the distinct impression that we were all in the body of a large serpent or lizard. When alerted to sit up and participate with the group in chanting, I perceived visual images of large felines, jaguars, and panthers, as well as snakes, lizards, and crocodiles. I was reminded of the mummified crocodiles representing the god Sobek that I had seen at the ruins of Kawm Umbú, during a trip to Egypt in 1990.

As I again closed off the outer world, I began to pursue with focused tenacity my intention of communication with the former life. I began to have images of riding upon a black stallion in a procession at the coronation of Queen Elizabeth. The scene changed and I found myself at Mortlake outside the home of Dr. John Dee, who was my teacher in that prior life. We were looking into an obsidian crystal, along with Queen Elizabeth and some of her entourage. The scene changed again as I focused my attention on attempting to communicate directly with "my" former personality (Robert Dudley), in a direct face-to-face encounter.

I found myself in a *psychomanteum,* a place specially constructed for divination and communication with deceased relatives. There was a large mirror on the wall, in which a face took shape and became three-dimensional. This face and the eyes told me what I wanted to know even though the exchange did not last long. However there was sufficient communication to inform me that I had already in effect rediscovered the mathematical key that I, as Robert Dudley, had previously worked on under the tutelage of the Elizabethan magus Dr. John Dee.

Ayahuasca was now about to teach me the lesson that I had not consciously sought. I began to see a series of shortcomings, not only in the previous life as Robert Dudley, but now again repeated in this life. I saw the machinations of the ego-personality and its subtle deceit of the Self, of the true Monad. I realized that intellectual success is not the end-all, but rather love and compassion wisely applied. I began to sense and feel the importance of my family, my parents, and particularly my love for my wife. I was raised up to the summit. I was at the feet of Plato's Indefinite Dyad, the primordial duality or ultimate yin/yang, about to be merged into the One.

Then I began to feel as if I was in the United Nations General Assembly where the ultimate union of male and female was about to take place. The yang and yin were about to merge into the One. I became deeply concerned that the world would not be here tomorrow if I went one iota further. Curiously, the following day during the integration circle, when we recounted our experiences, just as I was saying, "I feared that the world would not be here tomorrow," one of the glass votive candleholders on the altar shattered with a loud bang.

As I descended from the height of the summit experience, I felt remorse over the games I and others play. I saw the dance of lives with the perpetual repetition of mistakes, deceit, and cruelty toward Earth and all her creatures. I knew that I was in a chamber of initiation, about to go through a death experience. I felt great sadness for the pain I had inflicted on others, whether intentional or otherwise. I began to hear the "swoosh" of a large snake. I felt my abdomen crackle as if the skin of a snake were being shed. As I relived the pain of mistake and deceit, I physically purged into the bowl. I knew I had died without really having to die. I had truly experienced a death and rebirth.

Images, thoughts, and realizations danced inside my head. I was gently brought back by our guide to the group sitting in the circle. We then participated in very meaningful, though at the time somewhat difficult, exercises of the voice. We attempted to convey some of the meaning of our experiences through vocal tones and chants. It was incredible. I saw our guide literally transform into an old Indian shaman. We closed with a thanksgiving ritual, and I eventually drifted off to sleep.

When I awoke at dawn, I felt compelled to return outside to the mountainous terrain. Each step I took was sacred. The ground upon which I walked was sacred. I climbed as high as I could onto a precipice so as to contemplate the sunrise. I asked for forgiveness from the Earth for any harm I had committed. I studied the rock formations and found one that appeared to be a lizard with a human face. Fortunately I had grabbed my camera on the way out, and captured the incredible image. The Earth was speaking to me. I had acquired the meaning I had sought in the circle session, and so much more. The spirit of ayahuasca taught me that all is alive, all is sacred, and all is ultimately love-wisdom.

In retrospect several years later, I see that this experience opened many doors of perception for me. It allowed me to confirm the insights into

Pythagorean mathematics that have been the intellectual focus of my present life, through seeing its origin in a previous life in Elizabethan England. The recollection of this lifetime also helped me to understand the roots of certain other interests and proclivities. For example, my instinctive love of horses in this life feels closely parallel to my previous experience as Master of the Horse for the Queen. My present life's efforts and successes at tennis, racquetball, and athletics were foreshadowed on the tiltyards and tennis courts and in the forests of sixteenth-century England. My continuing interest in alchemy, sacred geometry, and the hermetic sciences are rooted in my previous work and study with the mathematician, alchemist, and magician, Dr. John Dee. Habits and frustrations of that life continue to be paralleled in events of this life. The real beauty of the teachings of the ayahuasca spirits is that they can help provide meaning, purpose, and direction to one's life.

In subsequent work with this particular plant teacher, I have been able to access other previous lives and their significance for my present journey. Ayahuasca has allowed me to explore the deepest philosophical issues of the self. I have been given insight and remembrance into the thorny issues and dilemmas involved in the Hindu doctrine of *atman* ("Self") and the Buddhist teachings of *anatta* ("Not-Self"). I have come to understand that these are difficult concepts, not because of the underlying reality, but because of the limitations of our conceptual formations and linguistic expressions. I have learned that by posing the questions appropriately, the plant teachers will assist with the response.

As a result of my ayahuasca experiences, I now have a greater clarity of direction in my life, a recognition of the sacredness of all living beings, and the presence of consciousness at all levels of organization within and throughout the cosmos. The ayahuasca plant teacher, much like the entheogens probably employed in the Eleusinian and other ancient mystery religions, assists in the sought-after remembrance, what Plato referred to as *anamnesis*. Our birth truly is a forgetting, as the poet Wordsworth said. Thank heavens there are techniques that can initiate one into deeper states of remembrance and recollection. The ayahuasca plant teacher has helped me to begin to answer, for myself in my own small way, the perennial questions: Where do we come from? Why are we here? Where are we going?

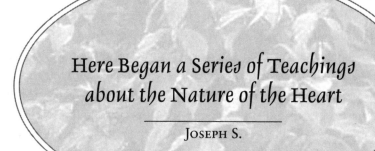

Here Began a Series of Teachings about the Nature of the Heart

JOSEPH S.

A writer and consultant in his forties describes initiatory ayahuasca sessions in which he obtained psychological insights, confirmed knowledge previously acquired through channeled teachings, and embarked on an intensive program of training with ayahuasqueros in Peru.

On this my first experience with ayahuasca my intentions were to prepare for my fortieth birthday, to allow more essence to flow through me, to increase my humor level, and to heal my right shoulder blade, which had been severely bruised in a fall off a ladder.

Visions began with some animal impressions after about forty minutes or so. My face became a tiger or large cat. Visions progressed rapidly with strong psychological content. Each emotion seemed very intensified. Some difficulty keeping track of any one thought. Just a stream of consciousness. At first some mild physical discomfort in stomach. Awareness of teachers showing me aspects of my persona and shadow sides.

My own narcissism or arrogance was a preoccupation for a while. First I felt greatness and wonderfulness as a special type of person with a significant mission in life. Then I became concerned that I was feeling overly self-important. Self-deprecation began to set in. I struggled between these two points of view, trying to know what the real truth was for what seemed like a long time.

Perceptions of my children. Great love for my son, C., aged 5. Great pain perceiving my relationship with my daughter, A. Felt that I wanted to be closer to her, but that I see so much of myself in her that I have a hard time relaxing with her. Missed my wife and felt like I wanted her to be with me. Had a strong vision of snuggling with her and the children in a big pile like kittens. Very happy with this. Smelled their animal smells, very familiar.

Aware of feeling isolated and difficulty letting love in from others. Felt that I must be a difficult person for others to give to. Struggled with this for a time.

Some illness set in after thoughts about the horror of the violence in Ireland. Got a strong sense of the law of karma and the awfulness of people feeling so separate from one another that they could kill each other. Sensed the horrible karmic consequences of these acts. Vomited and felt better. Sensed how karma can be greatly mitigated by forgiveness and surrender. Prior to this I fought off the impulse to vomit, thinking it was embarrassing in front of other people. Struggled with my own arrogance and wished to appear okay to others. Realized that I would have to vomit to overcome this.

I had been doing a lot of work with the seven roles or character types, according to the Michael teachings, and their specialties on the planet. *Artisans* and *sages* are expressive types; *servers* and *priests* inspiration types; *warriors* and *kings* the action types; and *scholars* the assimilation type. Became aware of fun, and how sages bring fun into the world. I felt the sage in me and experienced lots of mirth. I had visions of sages I know with expressions of intense fun on their faces. Then I thought about priests and how good they are at forgiving people. We give them the power and authority to forgive and heal us. Their specialty is compassion. Thought of the service of servers, the grandeur of kings. Had a vision of my son, C., a king, sitting on an emerald green throne with a huge emerald suspended over his head. Brilliant cartoonlike visions accompanied these perceptions. Lots of forest greens and strange looking creatures that seemed to be included in tapestries or huge suspended paintings, slightly grotesque and not animated but fixed.

Saw myself as a beacon, as a light that could be sought out by others who wanted to pursue knowledge that I might have to offer them. Perception of a fraternity of more aware beings that could be emulated for their integrity, wisdom, and overall way of being. This fraternity of older, wiser souls could transform society if they became the models for others rather than the violent stereotypes on TV.

Perceptions of my childhood and my relationship with my mother. Experienced my fear as a child. Indescribable how fearful I was of my mother at times. Recalled her being quite crazy for a time—violent and unpredictable. Felt repelled by her refusal to look at herself in this lifetime. Saw clearly how much denial she has been in. Saw her as old and sad and cowering. Then felt compassion for her. The need for forgiveness. The need to stop judging her and be understanding. Felt bad about my neglect of her and dad in their old age.

Looked out the large window in the living room toward the outside. Saw a bright light in the sky, moving slowly and without blinking. Got the distinct impression that it was not a plane. Thought it could be a UFO. Had never seen one before and got quite excited. Again, wanted to tell others but it was too difficult to communicate. Visions increased manifold after this.

Wondered whether there was a ceiling or limitation to this drug that gave it its strangeness. Is this a higher reality or is this a drug version of a higher reality that is severely restricted by the nature of the drug? Would my experience be the same if I took it again? At this point I became aware of others in the room. I noticed someone doing *kriyas* [yogic energy discharges—Ed.] and he seemed rather ecstatic. Wished I could feel ecstatic too.

I became aware of healing properties. Intensely aware that love heals all. I enjoyed feeling the rapture of loving between people. Then I worried that I was simply thinking and not experiencing. I realized how much I tend to do this and suffered for awhile under this perception.

I became aware of the nature of illusions. The difficulty and impossibility of paradox because each side is offset by the opposite position.

Paradox demands the simultaneous holding of both sides of duality. The only way to meet it is right down the middle.

Body vibrations. Streams of vibrations shaking my entire body at various times during the evening. These were not particularly ecstatic, nor were they painful. I witnessed them as somatic releases, perhaps *kriyas* that I had seen and wished for earlier on in the session.

Perceptions of ayahuasca spirit helpers circling around and around in groups of three. Each was smiling broadly and had three colorful feathers sticking out of a headband, very large eyes. They were full of humor and mirth.

About seven months later, I had my next ayahuasca session. The first image was of a small monkey, spry and cute, swinging with its family through the high limbs of trees. It communicated to me that it was good at getting into small places.

When the visions became intense, I noticed that I became skeptical and doubted the visions. I saw that visions in general are creations of artisans on the astral plane. I also saw that the artistic and creative artisans are a part of my larger self in the Tao. I saw their delight in working with images and forms and colors. I remembered not to get caught up in these visions but to press on and see if I could reach another level of awareness. I thought of my guide and teacher Michael and grabbed a piece of sugilite, a small purple rock from Africa, and another rock that I had brought with me, to help me get in contact with him better. I held the rocks in each of my hands and felt their effect as they sent energy up my hands and arms into my torso, opening up my chakras to become more receptive. I realized suddenly that the ritual substance was the main teacher in this event and that bringing in another teacher was not necessary. I put the rocks down.

I wondered if I should lose myself in the pleasure and enjoyment of the visions or push on toward deeper and more profound insights about the nature of consciousness. After a while I relaxed about this and the visions continued, but I felt I had gone deeper somehow.

I felt the softness of my heart. I felt indescribable tenderness. The moment I focused on my heart, everything became quiet. The whole surrounding universe seemed to pay respect and became peaceful. Here began a series of teachings about the nature of the heart. I learned that this is the center for essence contact. Love is its vehicle and product. It is a sender and receiver of love. Love is incredibly powerful. Love is the food source of life for all beings. Children, animals, and plants gravitate to it and feed off of it. They bask in it. A little bit will go a long way. All it takes is a fraction to make a big difference. As we experience love, we can then give it. When we give it, we instantly get back an outpouring of love-energy from animals, plants, and elements. I felt loved by the universe. I realized that I could love myself.

I saw that my mother and father wanted to fully express their love for me, but they were frustrated by their own challenges and limitations. I allowed them to love me, and I loved them back. I saw that they got me going in this lifetime. That they gave me just enough love to launch me. I saw that I became an expert at making a little bit go a long way. I heard my guide Michael say "Reach out for the higher centers," and as I did that, I experienced more love feelings.

I looked out the window and saw Grandmother Ocean. I saw the trees, the clouds, the Earth, their properties and principles. I saw their life. I loved them and they loved me back. I saw the power of life coming from the earth. I felt myself ball up and become the earth. I saw the clouds and laughed to see the pattern there. I saw South American images of the gods there. I saw the ayahuasca spirits in the patterns of the clouds. Was this a metaphor? Was I projecting these images? Did it matter? The fact is, I was seeing them. They kept moving together but never colliding.

Eventually I was pulled away from the window by the circle check-in. I found myself able to listen and speak from the heart. I spoke openly and freely without fear about my experience. I felt tremendous compassion for the woman next to me who was so sick. I knew it was her problem, though, to work out, and I could stay separate but understanding.

I felt much relaxation of my shoulders after I learned to work with the energy there. I could approach my shoulders from my heart and send them love, and they would respond to that kind of healing. I understood how hypnosis works. It lifts the binding, blind responses of certain defenses or behaviors and gives them permission to stop. They are commanded to stop by a higher authority. They need a directive to give them their new instructions. I saw that whatever I said as a command would take effect during hypnosis.

Looking back several years later, I see that these first experiences with ayahuasca have had a powerful influence on my life over the long term. They led me to many further explorations with this plant teacher through the assistance of an ayahuasquero in the jungles of Peru. I cannot begin to explain the many insights into myself and others and the liberation these visions have given me over the intervening years. In fact I am now working on a book based on a personal account of these powerful experiences and events.

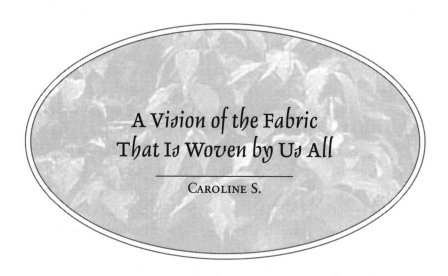

A Vision of the Fabric
That Is Woven by Us All

Caroline S.

A meditation teacher and artist in her thirties describes how in her ayahuasca experience she spontaneously exorcised a toxic psychic entity by purging it, and then tuned into the supportive web of life and spirits woven by women and men, and indigenous and modern peoples.

During my first experience with ayahuasca, which occurred in a private residence in northern California, I saw an image of the Great Spiral and had the physical sense of moving within it. The images, colors, and depth intensified until I was totally within the Spiral. I felt a knowing that God was there, teaching me my place on the Spiral. Once I understood that, the faces of all the people I love in my life began to appear and move into their place upon/within the Spiral. I had an incredible sense that everything was perfect right where it was.

I suddenly became aware of one of the men in the circle becoming very sick, and I knew it was because he was struggling against the force and momentum of the Spiral. I telepathically told him to "relax, move with it," as I astrally moved into his space to support and guide him. He continued to retch and struggle for a few moments and then became aware of my presence. Others in the group had also sent their energy and/or astral body for support. He then relaxed and began to move forward and upward with the Spiral.

The Spiral changed into an incredible area of third/fourth dimensional energy. Colors were deep, and everything had movement, life, and presence. I could feel form and creation and my presence in it. From the diamond pattern emerged the sacred serpent, a boa, who taught me about the importance of the Spiral, the kundalini power/movement in my own body, the journey back to God.

I had a feeling of needing to urinate and stood up to go to the bathroom. As I stood up, I became aware of the Plant Kingdom Guardians sitting along the side of the room. They were in deep meditation, and I could sense that they were overseeing our visions. I felt that they were permeating the room with their presence, giving me a sense of respect and knowing of Plants that I had never known before.

I was then assisted upstairs to the bathroom. While sitting on the toilet, I suddenly and violently had an intestinal spasm, while at the same time receiving an image of a dark hole with the diamond patterns. Up from this dark hole sped a very powerful and scary demon, which I *knew* did not belong within my being. As I vomited, the demon rushed out of my mouth. Everything went blood-red—the sink was filled with blood and my hands were covered with blood. I was not afraid, for I knew that this demon had gone to its right place. As I let the image move away, the blood turned to water and then was gone.

My power animals, Jaguar and Mountain Lion, came to teach me about physical strength, about breathing through my nose with my mouth open to smell and sense my surroundings. I had a great sense of feminine strength, knowing that the clan depends so much on Her wisdom and abilities. I had a realization of the need for women to recognize their power, to claim their rightful place within the clan. I know the importance of men and women respecting and supporting one another's power and place, but this gave me a sense of the urgency of the Great Mother for women now to come forward and to reset the balance.

I saw how the women's liberation movement was pushing too hard in the opposite direction and was not in its proper place on the Spiral. I experienced myself among my people (Cherokee) in a tribal/clan situation where the women got together and built power together; they were extremely strong and purposeful. Yet, they did not seem to make their actions and purpose obvious to the tribe. They generated an incredible presence, a fine thread of consistent power out into the clan, weaving the fabric of clan society with their imagery and heart energy.

I then got a vision and physical sense of the fabric that is woven by us all. I could see the fabric being woven by our circle as we all focused on our own and the Earth's healing. I could sense each person's color and texture of thread, and how it all contributed to the fabric. I was then shown my meditation group, and how that fabric is getting so thick and strong and beautiful because we meet every Thursday, adding energy and material to the fabric. I knew the importance of families, healing groups, coworkers, etc., being aware of this fabric, and consciously building and creating it. This is how our world is created.

As the visions became less intense, I became very aware of the sense of family/clan among the circle members. It was obvious that, although we had each had different experiences, there was a very strong and stable presence of commonality and support. After the circle was closed, some of us moved out to the hot tub or sauna, continuing to share our experiences. In the morning, we shared our stories. It was amazing how interconnected the teachings were. Some people had experiences of Nazis and Jews, others of Native American Indians. There was a beautiful sense of belonging, and we all felt that much healing and teaching had taken place—both individual and planetary.

I felt honored by and grateful for this powerful experience and know that it has had a significant and positive impact on my knowledge of my Self, my place in the universe, and my purpose on the Rainbow Path.

Ethereal Serpents
Held Me in Thrall

SHYLOH RAVENSWOOD

In this account, a psychiatrist in his midforties relates a dramatic tale of how the ayahuasca snakes helped him purge toxic residues from unhealthy behavior patterns in his life, thereby gaining deeper self-understanding.

After a nine-month break from medicine circle work, I felt ready for another journey to the shaman's world of power animals and other inhabitants of the deep unconscious. I also hoped to reconnect with my physical body, from which I tend to become alienated by my writing, healing vocation, and other intellectual pursuits. Over years of personal experience with psychedelics, I have found that the structured ritual of a medicine circle provides a safe environment for such inner travel, and that the surest way to achieve this dramatic shift of consciousness was with ayahuasca.

I had ingested that Amazonian plant spirit helper on three previous occasions within the context of a medicine circle and had found the visions to be glorious. Mostly I saw entrancing swirls of multicolored lights intermixed with shimmering, kaleidoscopic patterns that overwhelmed my senses, creating a state of blissful absorption. Occasionally I caught a fleeting glimpse of a large cat or a raven—animals with which I feel a special connection—but they had seemed uninterested in me and did not linger in my awareness.

During the several days that preceded the circle, I had noticed tight, sore areas in my shoulders and neck, energy blockages that I attributed to my tension-filled days with acutely ill psychiatric patients. In my chronic state of having more-to-do-than-time-to-do-it, I had been neglecting the good dietary habits that I usually maintain. Arriving home late and tired, I often unwound with a bit more alcohol than was my custom, which diverted me from performing an essential inner review of that day's emotional residue that promotes healthy sleep.

During the afternoon vision quest in the San Gabriel mountains that preceded the evening's circle, I tried to concentrate on my intention for the weekend, but I allowed myself to be distracted by a series of mundane thoughts that blurred my focus. In the hour before the circle began, I again avoided turning inward by chatting idly with my fellow travelers as I waited my turn for the shower. I realized I was a bit constipated, but in the excitement that precedes medicine work, I neglected to relieve myself before the opening ritual.

It is for good reason that Amazonian natives refer to ayahuasca as *la purga*. My past experience with this mysterious medicine included gagging robustly as I swallowed the revolting brew, followed by an inevitable episode of vomiting that precedes the peak of the experience. Although I would rather forgo that aspect of the journey, it had seemed merely incidental, a minor toll for passage into a world of extraordinary compelling visions and deep psychological insight.

But this time was different. The thick, pungent medicine seemed especially foul, coating my tongue and palate, sticking to my throat as I tried to gulp it down. Once in my stomach, it felt as if I had swallowed a live boa who was inching through the acidic labyrinth of my guts, pausing to squeeze them tight in sequential spasms. Barf bowl at the ready, I repeatedly leaned over to expel the herpetic brew, but my retching yielded nothing. As the early visions mounted, I was preoccupied by the incessant rumblings in my abdomen and my mounting frustration as I repeatedly failed to rid myself of that relentless reptile.

Now I realized I was getting into a bit of trouble. Nearly overwhelmed by a swirling visionary wash, I could feel my pulse quicken and blood pressure drop as blood rushed to my heaving intestines and away from my brain. I knew I wouldn't lose consciousness as long as I remained supine, but I also knew that I should sit up and try again to vomit, which caused me to feel light-headed and weak. As time passed, my predicament deepened, finally brought to a head by a sudden, compelling urge to evacuate my bowels. After lurching to the bathroom, which was lit by a single candle, I sat and lowered my head between my knees.

Snakes! The small room was suddenly full of writhing serpents, crawling on the walls, ceiling, floor, over my body, into my nose and ears. Some of them morphed into other animals—spiders, wolves, fish, fabulous multicolored firebirds—then abruptly reaffirmed their true nature as snakes, hissing in my face with fangs bared and tongues flicking. I was sure they were laughing at me, mocking my discomfort.

In ordinary life, I do not fear snakes but admire and respect them as fascinating and mysterious inhabitants of the California chaparral where I jog daily. And now I was not repulsed by these reptiles either but felt enchanted by their momentous power and presence, curious as to what they wanted with me. Later, the shamanic guide who led the circle suggested that if I had reached out and captured one, it would have taught me what it knows. Perhaps next time. . . .

Finally after achieving explosive relief from both ends of my beleaguered alimentary tube, I was at last able to sit up without feeling woozy. The circle's sitter came to check on me and helped me back to the group, which was just starting the first round with the singing staff. When the staff came around to me, I weakly propped myself up on my elbows and hummed a barfy song to express my ordeal. The peak of the visions now past, I spent the rest of the evening reflecting upon the meaning of this exotic visitation, trying to gain a teaching that seemed as slippery and elusive as those ethereal serpents who momentarily held me in thrall.

Only in the weeks that followed did the word "obtunded" arise from that night's events. This word signifies a reduction, stubborn blockage, an impediment to the natural flow, a dulling of the senses, emotions, and intellect, a constipation of spirit. I had allowed myself to become obtunded in many subtle and not-so-subtle ways. The snakes were there to open my heart, beginning with my stupidly neglected guts.

Now, I won't say that I no longer get obtunded, that I steadfastly heed my body's subliminal signals, or that the shallow cravings of ego never impede the flow of Spirit through my heart—only that I am a bit more aware of my lapses. Maybe when I catch that snake. . . .

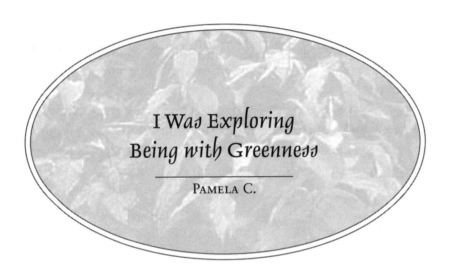

I Was Exploring
Being with Greenness

PAMELA C.

A psychotherapist in her forties overcomes her fear and repulsion toward Earth and nature, grieves for the Earth, delights in letting ayahuasca elves heal her body, and rejoices in brotherly relationships in a circle of men.

My purpose for the ayahuasca session was to explore and heal my relationship with the Earth. I have been repulsed and fearful of working in the dirt, or taking walks, or being in nature for many years, ever since I was a child.

The liquid substance tasted awful and compelling at the same time. I laid down and waited, with my bowl close by, for the vomiting I had been told to expect. When it came I wasn't expecting it. No horrible cramps or inner need or awareness that I needed to vomit, as with an illness. I simply opened my mouth and my insides came out into a silver bowl. It was all very gentle, almost delicate. I felt feminine and dainty, not wretched.

I was lost in inner exploration when I noticed I was crying and full of grief. Grief for the world, and it came to me that it would be helpful and healing if women wept for the Earth. I thought I must be authentic, I must weep with my clients. Especially my women clients who are HIV-positive.

Later, just going with the experience, I was exploring being with greenness and deep, lush vegetation. Then I noticed little cartoon people in pastel shades of pink and blue. Not at all the earthy green folks I had expected, if I were to create them. They were working on my body with little tools. They looked up and were happy to see me. It was as if a princess or queen of the land had come to visit. They were pleasantly surprised and proud of themselves and their work. They told me not to be so concerned with fixing my body myself. They would handle all repairs. If I wanted to do something, then before I went to bed I could focus in on what part of my body I wanted healed, but not to do it myself. I felt much relief, and I trusted them and their workmanship. An image of a sloth came to mind—one of my favorite animals—with the saying "Just hang in there."

Again, back into the jungle, just letting myself wander and wonder.

Noticed that I was on my knees arching my back then letting it fall, like in hatha yoga positions. I was not intellectually directing this process. My body was moving in a way of restoring balance and harmony to my back. I remember the guide asking if I was all right, if I needed anything. "No," I answered him. I was simply moving my body in a way that would heal my back. I remember later feeling extremely comfortable in my body. I was elongated and felt sensual, not sexual.

More awareness of moving through the jungle. The jungle moving aside as I quickly and stealthily, with focus, passed. I was the jaguar. I saw the jaguar's face and the beautiful wondrous anaconda as my journey came to an end.

Coming back to this reality, I felt surrounded by my brothers. I was the only woman in the group with about eight men. I felt such joy in being amongst them. No sexual energy from them, but a sense of brother- and sisterhood. I felt grateful for the brothers of my family and these new brothers. I was seeing men in healthy relationships with other men. I felt safe, that all was right with the world. I felt protected by them and safe.

In the months since the session, my relationship with earth changed dramatically. I started a garden and still work in it. Every time I work with the earth or with plants, I learn a deep truth about life: weeding, planting, fertilizing, preparing the soil, treating plants individually, each type having specific requirements, just reflecting about relationships and that some are annuals and others perennials. I walk, generally by myself—four or five days a week. A miracle for me. I feel the insides of my body more deeply. I have reverence for insects that is new also, a deeper love for all animals and birds.

Finally, and this was a big surprise, my fear of heights has begun to fade. I can go over the San Rafael bridge without trembling or breaking out into a sweat.

I have had the deepest and longest lasting results from this medicine and give thanks to it and its teachers.

Reviewing this experience from the perspective of several years later, most of the lessons I learned in the session are still with me. Sadly, I must report that the fear of heights is ongoing. However, my feeling of safety in nature continues to grow. I have even taken people out into the wilderness for retreats, something I'm sure I would not have been able to do without the help of this medicine's teachings. My dog has become my constant companion, whereas in the past pets have always been an added chore for me. Now I find solace and friendship in the animal kingdom. My treatment of insects continues to be respectful. I avoid all use of pesticides in my house or on my food.

Recently, in a book on nutrition, I found references to *eicosanoids* —super hormones that "control every vital physiological function" and are "the most powerful biological agents known to man." They are found in all living things and have been unchanged for millions of years. Rereading my ayahuasca account, I connected the little pink and blue healing workers I saw with the eicosanoids in my body. I was glad to be reminded that they are taking care of my body and that I need do very little to help, though attention is always appreciated.

After the ayahuasca session I began to rethink my relationship with my brothers. I had spent most of my time in psychotherapy focusing on my parents. I now began to see the lack of relationship I had with my brothers and have made changes in my life to have more connections with them. This has been a difficult but enriching experience. The ayahuasca session gave me a touchstone for what a healthy sibling relationship could be.

Liquid Plum'r
for the Soul

I. M. LOVETREE

An educator in his fifties relates deep purging and healing experiences with ayahuasca, confirmations of Buddhist teachings, and important understandings of the tricks of the ego.

My intention was to reconnect with the medicine way of healing. Fifteen years had passed since my last psychedelic venture. I had soured on the whole New Age trip, having been badly burned by a meditation school to whom I had given my all, only to wake up one day to find that the school wasn't what it was cracked up to be. There had been hints all along of corruption, hints that I had consciously chosen to ignore in the interest of giving myself over to a teaching and in the interest of retaining my membership in a spiritual community for which I had made many personal sacrifices.

I both left and was kicked out of the school. My departure had something to do with my not respecting their authority. I was also disenchanted. I mean this literally. One day, during a meditative training session, I became disenchanted. The spell I was under simply fell away. Eventually, I returned to my original path of medicine work.

First Session

I lay pinned to the ground, flattened, possessed. I had done medicines before but they were nothing like this. The ayahuasca spirit vibrated and roared. Like the Santa Ana winds winding their way through mountain passes before their window-rattling, tree-shaking descent, I could sense the ayahuasca coming before it arrived. Here it comes again! A huge serpentine roto-rooter moving through my systems, sparing nothing, unearthing everything. I did not throw up or vomit. I puked, yes, puked. Everything came up. Whatever I couldn't stomach was dredged up from my organs, from my cell beds, from the very depths of my being; *la purga* was at work. Resistances, defilements, long-held, hardened-with-age resentments, all bejeweled and resplendent, cascaded into the bowl set before me. Interestingly, I enjoyed the process. I felt instantly relieved. The wave of nausea that had preceded and triggered the upheaval quickly subsided. The spectacular display of brilliant DMT-induced visions brightened and sharpened as I settled back to await and experience that which lay before me.

Scenes from my life flashed before me, revealing a startling array of weblike interconnections, a particular destiny playing itself out over time. Seemingly separate unrelated events appeared to be inextricably interrelated, like my relationship with a Navy buddy some forty-one years before. I had never understood his alcoholism or why he selected me to be his shipmate, bunk buddy, and pal. I now saw how our destinies and our paths were intertwined, how similar we were. I also saw how difficult it was for him to be a young, fresh-off-the-reservation, far-from-home, full-blooded red man stationed in the deep South before the official color boundaries had fallen. So much of what I do now, my organic farming, my environmental activism, my medicine work, and my counseling practice have, in some mysterious way, been touched and influenced by this relationship that, as it has turned out, was to be the first of many such involvements with the teachings and the ways of the First People, of this and other lands.

Most Recent Session

Since then, I have participated in seven or eight ayahuasca sessions. Each one different from the others, distinctively different. It's always that way with all of the medicines I've experienced. When it comes to medicine work, set and setting are everything. Intention is the key. Whatever arises in the vision quest or in the circle will in some way, however tangentially, be aligned with one's intention. My most recent ayahuasca session proved to be no exception to this rule.

My body had been acting up: there were signs and symptoms galore. I was tense, agitated, stressed to the max. All of my problems seemed to be coalescing: my work scene, my home life, even my reason for being, all were in a state of agitation. It was clearly time for a spring cleaning, time to ride the green tide of rejuvenation and renewal, and yet I hesitated. I feared this would be another of my working sessions, a dredging up of all the muck and stuff I had been avoiding and storing. In short, all of my image formations, my usual holding patterns were fully activated and ready to be processed. Everything in me cried out for ayahuasca, for *la purga*, and still I hesitated.

If you have ever tasted ayahuasca you will understand why. Even the sight of it evoked a shiver. Getting ayahuasca down is one thing, holding it down another. Lying still helps, as does breathing into the nausea, but nausea wasn't the issue with me this time. It was fear. I couldn't fool anyone, least of all myself. Talk about approach-avoidance behavior. One part of me wanted to do this, needed to do this, another part of me resisted at all costs and was busy looking for an out. Something happened during the precircle desert vision quest that helped me to decide.

We had been told by our guide that the appearance of an animal spirit was especially significant when it appeared four or more times. Although lizards were to be expected in the desert this time of the year, I had seen at least a dozen of them as I walked up an arroyo. Something happened that seemed very significant. Two large lizards measuring perhaps sixteen inches in length popped out of the desert shrub not more than ten feet from me onto the pathway on which I was walking. The lead lizard looked at me, as did the other, and then they both scampered back into the shrub only to return again and again to repeat the process, each time stopping to look at me. They were playing. They were also performing—for me! In my mind, lizards are associated with snakes, and snakes are associated with the serpentine ayahuasca spirit.

The choice was clear and the results of that choice, as it turned out, couldn't have been better.

I drank the ayahuasca brew and settled in for the night to see what magic it would bring. Over time and with practice, I had learned to cut a deal with the spirit of the medicine, and with my Self. Simply put, I would humbly implore the spirits of the medicine to be gentle with me. Of course, dosage had a lot to do with it. Too much medicine too soon proved to be an inappropriate approach. My ego defenses would be overwhelmed leading to reflexive grasping, to desperate clutching, to panic. Wait-and-see-and-add-a-little-booster-later-on proved to be much better strategy. What's the point of going in if you can't navigate? The combination of appropriate dosage coupled with an attitude of mercy toward myself was the way to go.

The guide suggested we use the acronym *b-a-l-e* [breath, awareness, light, empathy—Ed.] as a navigational and orienting device to help us stay present and on track during the session. The first two, *breath* and *awareness,* were relatively easy to maintain, and having spent years in a school that worked with light-fire energy, I was used to working with *light.* I had also studied the Buddhist teaching concerning *bodhicitta,* which refers both to an awakening state of mind and to a practice for developing compassion. Yet there was something about having empathy for myself that really struck home. Although in recent years I felt I had become less judgmental about others, I still tended to be hard on myself.

In a flash of insight, it occurred to me that if Buddhism could be reduced to a single concept, it would be empathy or compassion—for oneself and for others. Still, from this place of heightened awareness, I became conscious of the seemingly endless habitual flow of judgments. Each judgment, in its moment of execution, took me out of presence and put me back into suffering. I thought about the Tarot card Judgment, on which trumpeting cherubs are shown celebrating the liberation of one who has triumphed over judgment-bound consciousness.

I also noticed how my thoughts would steal me away. One moment I was at one with the music we were playing, the next I was commenting to myself about it. There was a world of difference between the two. I saw, with crystalline clarity, the subtle and pervasive interlocking habits of mind that collectively constituted my identity, some of which were derived from parental habit-of-mind streams that had converged and were alive in me. I understood the meaning of the Basque prayer for the newly born, "May this one be the one to break the harmful family patterns." I felt as if I were riding through time on the tip of an evolutionary spear, that I was that tip. I knew that a change in me would, in that moment, initiate a change that would reverberate throughout the web of life. I saw that the sins of the fathers were sins of habit that could be transformed and transmuted, and I realized I was here to do just that.

I had asked about eldering, what I needed to do to become a better and wiser elder. The answer I received was direct, immediate, and brief: "Continue to work on your self, and the rest will take care of itself." I saw that under so-called normal conditions, my ego shoulders Spirit out of the way and claims all territories as its own; my ego simulates the self. The conditioned ego does this to defend itself against the perceived intrusions of Spirit, which it regards as threatening to its integrity and survival. I could see why Jung regarded mainstream religion as a defense mechanism against religious experience. Most religion is Spirit co-opted and made safe by ego for ego. No real sacrifice or transformation is required.

Various other tricks of the ego were revealed to me during the course of the ayahuasca medicine circle. I observed the rapidity with which my brain-mind, in the service of my ego, took my awareness out of the immediacy of the moment. The workings of the conditioned brain-mind were subtle, elusive, and quick. The transition from being an active participant in the life flow to being out of it could occur in a fraction of a second. I also observed a profound change in one of the other participants who I sometimes found to be overbearing and unavailable, sealed up and trapped behind a professional facade. I have explored this territory, I know what it is like to cling to an ego-driven prison of one's own making, what separation and suffering the maintenance of such a position entails. I was moved to tears as I watched the medicine he had taken loosen the grip of his ego, allowing spirit to flow and express itself, rendering him receptive, vulnerable, and true. It was beautiful and heartrending to hear him cry out for mercy and forgiveness, to hear him acknowledge his being caught, to see his tightly held boundaries dissolve.

As the effects of the medicine wore off, I found it fascinating and instructive to observe the process in which the constellation of conditioned images returned to reassert their claim. Like pieces in a jigsaw puzzle, the elements of personality-ego began their slow but inexorable slide back into place, re-creating a formidable interlocking facade, an iron mask from behind which the light of spirit could be but dimly perceived.

I have learned that personality images will rebound. It's as if they have a life of their own. However, the good news is that the sacred medicines also leave their imprint. Neural networks have been changed, rearranged, re-created. Healing and visioning have taken place. In some subtle way, my circle friend and I will never be the same. I have also found that the work of grounding the lessons learned in the circle takes place after the circle. The process continues. Opportunities to integrate into ordinary reality what one has learned in the session will arise. The effects of repeated ingestions of ayahuasca over time appear to be cumulative. For example, my sense of connection with the plant kingdom has been substantially strengthened. Increasingly, the Green Man archetype, with which I associate the spirit of ayahuasca, influences and expresses through me. My farm is now 100 percent organic; my commitment to environmental action is now taking a bigger bite of my time, money, and energy. There is an old Sufi admonition about not inviting an elephant trainer into your living room unless you are prepared to live with an elephant; this is especially true if the elephant is green.

Afterglow

With ayahuasca there is a pronounced afterglow, of which I am conscious of for two to three weeks, sometimes longer. I have a palpable sense that all is well with the world, a sense of being effortlessly at home in my body, of being fully and easily present. Emotional perturbations are likely to be more short-lived. Coming back to center seems to occur on its own. After the last session, I experienced a very subtle but persistent sense of spaciousness as if a refreshingly cool mentholated breeze were blowing through my body. I didn't have to summon it up. It was there. All I had to do was notice it.

Seven weeks after the medicine session I had an extraordinary dream in which my ego was like a fiddle being played by Spirit (Higher Self). Even more spectacular was the fully orchestrated symphony I heard. The music was crystal clear, ethereal, exquisitely wrought. Since I can neither read nor write music, I was left wondering who the composer was. I felt like a Beethoven or a Mozart walking around with spontaneously orchestrated symphonies in my head. I also realized, while still dreaming, that the dream was metaphorical.

After the ayahuasca sessions, I feel cleansed within, throughout, and all about. I have a sense of having been healed at all levels, especially the physical. The ayahuasca medicine seems to have a special affinity for the gastrointestinal system: it snakes its way through the body, seeking out and eliminating obstructions to life energy flow. I sometimes think of it as a form of kundalini, a Liquid Plum'r for the soul. For cleansing and healing, for reconnecting with the vegetable kingdom, ayahuasca is definitely my medicine of choice.

The Plant Spirits Help Me to Heal Myself and Others

EUGENIA G.

This is the story of a pharmacist in her forties, undergoing a physical and psychic healing crisis, or spiritual emergency, who through her experiences with ayahuasca, learned to communicate with the spirits of all kinds of plants in her everyday environment. As a result she became an herbalist-educator, undergoing a series of ayahuasca initiations in the Amazon, in which she learned to work with the "spirit doctors," like the traditional ayahuasqueros.

My purpose for this first ayahuasca session was to deepen my understanding of the changes occurring in my life at this time. A few years earlier I had begun to experience a spiritual emergence that seemed to occur spontaneously at unpredictable times. The phenomena of this spiritual emergence included physical disturbances (difficulty sleeping, excess energy, trembling, weight loss), psychic openings (spontaneous art and poetry, automatic writing), and heightened spiritual awareness (the sensations of two realities existing at the same time).

I began psychotherapy to gain a greater understanding of what was happening, to learn how to handle the energy, and how to open to these experiences and put them to good use. There were tremendous upheavals and spontaneous eruptions of energy. After gaining some understanding, knowledge, and control of these energies it became time to explore in a more focused fashion what these experiences had to teach. The ayahuasca circle setting offered the perfect opportunity. I was drawn to ayahuasca because I had heard of it being a catalyst for deep connecting powers and for grounding the energies of the lower chakras.

Like most of those who have experienced ayahuasca, I, too, experienced the breathtakingly brilliant visions of nature's plants, animals, birds, and river scenes. However, to my amazement, what happened during the ayahuasca experience is not as dramatic as what happened afterward. That's saying something, because ayahuasca is considered one of the most powerful visionary catalysts.

The morning after the ayahuasca experience, I noticed how unusually open I remained. I had the sensation that I was outdoors even while being inside the cabin. During a morning drumming ritual, I realized the plant kingdom was talking to me. Plants, berries, flowers, and trees were telling me of their medicinal properties and how to use them. The plant kingdom and I remained in communion all morning long, lost in our own private conversations. The plant kingdom had suddenly come alive to me.

I was astonished. The only green thing in my life at that time was a lone African violet sitting on my desk. The extent of my knowledge about plants was that they needed watering, period. Later that week while discussing the experience with my therapist, he suggested that I might be an herbalist-in-the-making. It wasn't so far-fetched, given pharmacy's historical background with botanicals. "You're joking," was my response, as I was busy with a conventional pharmacy career at the time. I had no interest whatsoever.

But, it seems, the plant kingdom had an interest in me. For months, I was haunted by dreams of plants, and how to use some of them medicinally when a loved one was ill. I tried to shake them off saying herbs were folk remedies, and I represented modern science.

Finally, following a friend's suggestion, I attended an evening lecture on herbs. Fascinated, I signed up for a group of seminars on herbs. I was hooked and committed myself to a year-long clinical herbalism course. The plant kingdom continued to direct me. I experienced a spontaneous healing from an herb in my garden. I had the good fortune to work with a physician who primarily treats herbally. I was being shown a new way to practice pharmacy from a natural medicine perspective and finally left conventional pharmacy.

It's been about four years since that life-changing experience. My work with the plant kingdom deepens. Plants and herbs have become my passion, a private and holy experience. They take me to the deepest parts of my being, my soul. After my first ayahuasca experience all kinds of plants began to speak to me directly of their medicine and how to use them. Wherever I went I seemed to merge with plants from the fields, gardens, houseplants, and even weeds. Ayahuasca first opened me up to the experience of the plant kingdom as a "way in"—a deep connection with the divine.

I have learned to connect with plant spirits through a variety of shamanic journeying techniques. The plant spirits help me to heal myself and others. I have come to realize my life's work: I am to follow the path of a shaman-healer, who does visionary journeying to help others. As was my original intention, I now have a focus, a technique, and a purpose that seems to be answering the questions raised by the spiritual emergence. The ayahuasca experience blessed me with a healing and an understanding beyond anything I could have imagined. I remain a humble servant and truly grateful to all who have helped along the way. My special thanks to the guide, the group, my therapist, ayahuasca, and the plant kingdom.

While I was becoming more deeply involved in the transition from pharmacist to herbalist, I had repeated nightmares in which I was tied to a stake with a large leather thong wrapped around my wrists. Repeatedly, I was burned at the stake for what I knew to be true, much as the witches were during the Holy Inquisition in the Middle Ages. I began to suffer from a disabling carpal tunnel syndrome, which made it impossible for me to perform my work functions as a pharmacist. I was forced to leave my original profession, as I had known it.

The repetitive dream finally completed itself in the following release sequence: as I was being burned at the stake once again, I cried out to my tormentors, "You can burn my body, but you can't burn my spirit. For I am back, and my name is now Eugenia. You will have to deal with me by that name, for I live." Suddenly, I was released from the burning stake. The dream did not recur. I left conventional pharmacy, and the plant kingdom provided me with a new profession as an herbalist-educator of botanical medicines.

Then ayahuasca called to my soul even more deeply. I began to feel it was time to visit that most primordial realm of healing plants, the Amazon rain forest. The overpowering size and expanse of the rainforest had a depth and density that I had never before experienced. Besides being rich with verdant fecundity and colorful wildlife, the rain forest holds secrets that could change the course of medicine as we know it. Even having worked with plant medicines for years, I was really unprepared for the magnitude and layered richness of the Amazon rain forest. The Garden of Eden does exist, and I was in the middle of it. I was at the site of natural creation, watching the ultimate masterpiece unfold before me. The rain forest's pure aliveness is uncluttered by our civilized neatness and what we consider to be the necessities of life.

Blown open by the sheer energy of the primal rain forest, I began to have unexpected spontaneous visions. Through a series of serendipitous events, I came to meet an herbalist-shaman who would eventually become my teacher. Shaken by these unsolicited visionary experiences, I asked for his explanation and help. He answered quite simply, "You have enough energy to be a shaman." Instantaneously we seemed to read each other's hearts and became as blood family. He agreed to take me on as his apprentice, and I agreed to follow his directions, including stringent dietary restrictions, to learn from his vast store of healing plant knowledge.

I had many visionary experiences with the shaman. I have seen the parallel worlds of spirit doctors in human and in plant forms. I have met the great feminine spirit of the all healing plants, the beautiful *Murcura*. She showed me a book called "The Rules of Life" and told me that when I needed help on behalf of myself or someone else, I was to come to her with the book, and she would show me the pages with the helpful answers on them. Ayahuasca has allowed my everyday life to come more alive—my skin became electric, and light was everywhere; lucid dreams, messenger birds, talking animals, and plant spirits continue to teach me.

The transformative powers of ayahuasca profoundly changed my life. I am apprenticing to become a shaman, learning from my teacher in the rain forests of Peru, and practicing my profession as an herbalist-educator in North America. To be a shaman is a way of life, not just a single or even series of events. Ayahuasca is one tool, a central key to the whole process of apprenticeship.

Sometimes I feel as though the ayahuasca plants are growing inside of me. I seem to be merging with the spirit of ayahuasca, which has a quite definite agenda. She reaches out her symbiotic vine arms to embrace those who come near her. My arms and hands have become the human arms and hands of ayahuasca. Its spirit is that of a great healer who helps all who imbibe her, to become who they really are. The rain forests are in my veins, in my blood; they are part of me, and I part of them. We are not separate. My life is no longer my own. Daily I let go into deeper aspects of ayahyasca, as my ego fades into the background of yesterday and the healer of tomorrow is born within me.

The Great
Serpentine Dance of Life

RAIMUNDO D.

A psychotherapist in his fifties experiences a series of progressively more radical and liberating purgings, as he confronts and becomes the great cosmic ayahuasca serpent.

My conscious intentions for this first experience with ayahuasca were to see what I needed to do to better serve people, and to meet the Serpent. During the waiting period just after ingestion the thought came up that I would have a frightening experience with a large snake, an anaconda or boa constrictor, and would be devoured or crushed. I decided I did not need this kind of experience and ended up not having it. I decided instead that I really wanted to *see* the serpent.

First-level purging: the inner message was that I cannot see the serpent because it is too vast—I had to *be* the serpent. Spontaneous undulations began as I became the serpent and experienced serpentine movements, swaying back and forth. An awareness of the processional quality of the human experience on the planet for aeons. The serpent was leading the human procession through all kinds of wondrous experiences. An endless/timeless experience of the great serpentine dance of all life. I caught a glimpse of iridescent, luminescent, semicircular structures and realized they were the scales and one curve of the serpent's body. Realized that all I could *see* was a relatively small arc in the great coiling circle of life's procession. The awareness that life is a "road show" on the side of the Great Serpent.

Second-level purging: the realization came that a cleansing/purifying/opening process was occurring, and that each vomiting took me deeper into the experience. I received a spontaneous visual showing of the difference between the plumed and the unplumed serpents, both limited versions of the Great Serpent. The plumed serpent is masculine, involves outer impression and show of power; the unplumed serpent is feminine, involving inner expression and statement of strength.

Further associations came regarding yang and yin and my own animus and anima. A very archetypal experience—at once cosmic and atomic. An abrupt shift into the physical structure of my own nervous system took place that I think was precipitated by the tape of the ayahuasquero chanting the *icaros*. I experienced my entire body being reprogrammed and rearranged, even reconstituted at the deep cellular level. This resulted in an incredible feeling of openness, solidity, and wholeness.

Third-level purging: this was the entry into the serpent's lair. The serpent had transformed into a kind of dragon. I never did see the dragon in its lair. The deeply felt sense was that a male and female dragon co-inhabited the lair. I had a visual and kinesthetic experience of being in a semidark cave, moist and hot, with floating steam puffs throughout. It was an extremely powerful place of creation and generativity. I knew the pair lived there, and it felt like a wonderful home to share.

Fourth-level purging: this occurred after I fell asleep and was primarily experienced as a complete cleansing of my entire alimentary tract and inner body cavity. Throughout the night and early morning, I experienced the cleansing moving through my stomach, intestinal tract and colon. An incredibly high colonic! By morning I experienced my entire body as a clean, clear, hollow, open tube. My throat, which tends to have a block, felt completely open, and my whole body felt totally clean and clear.

Regarding my intention to better serve people, I realized that (1) I needed to remember the vastness of total life experience and appreciate the limited arc or sector that is lived and experienced by me in this life-time; (2) I needed to keep my body as open, clean, and clear as I can to be a channel/vehicle/medium for visionary experience; and (3) I needed to *be* my experience of myself and others internally and somatically, rather than just observe externally and visually.

I was deeply grateful for having had such a rare opportunity to experience this experience. I thanked Mother Nature, the plant, the facilitator, the group, the sitter, and myself from a place of deep grati-tude, respect, humility, and awe.

The vivid reality and unequivocal truth of this experience has definitely been deeply internalized and has stayed with me for over a decade. The experience was an unforgettable one that has made a profound impact upon how consciously I live my life and how intimately I relate to and serve others, both personally and professionally. Through this experience, I feel that I was able to identify with the source of creativity and to undergo a radical awakening and self-transformation. This has resulted in what appears to be a permanent state of being that allows me to more openly and clearly receive, reflect, and respond to the real-ity, truth, and beauty of all living beings.

A Most Palpably Buddhist-like Experience

RENATA S.

A painter and experienced Buddhist meditator in her forties describes her one experience with ayahuasca, in which she reached a state of profound equanimity in the face of both the beauties and the horrors of existence.

My first, and as of yet, only experience with ayahuasca took place in a comfortable geodesic dome in a relatively rural setting. Along with the others in my circle, I spent the day fasting, meditating, and fine-tuning my intention—what it was that I hoped to investigate while on the medicine. This intention was dealt with very early on in my experience on the medicine. And although enlightening, it was in some ways the least important part of my experience.

What I would rather talk about is the less tangible, but perhaps more meaningful part of my experience. I had an overall sense of the exquisiteness of the journey—aware that it was very much a metaphor for the journey of my life. At times, I had terrifying hallucinations. However, I never was moved to pull back from these visions. Rather, I moved into them and through them with conviction, trust, and an appreciation of their own particular beauty.

As many of my generation in the 1970s, I was very interested in various Eastern philosophies. For quite some time I had been a vegetarian and practiced meditation and yoga on a daily basis. Still, I comprehended the various philosophical concepts and precepts of Eastern thought on an intellectual level only, and no doubt superficially at that. My experience with the ayahuasca (as was true of my experience with LSD) put me in touch with an understanding of these ideas experientially. It was as if my body accepted ideas of oneness, duality, paradox, etc., on a cellular level. The physicality of this experience was much more far-reaching than any intellectual understanding, and this made it one of the most palpably Buddhist-like experiences I have ever had.

I understood the sad and frightening visions to be every bit as wonderful as the most beautiful visions. The marvel was that I felt totally alive, open, responsive, and fearless! Accepting the fleeting nature of all, it was so simple to be fully present for every moment. Perhaps for the first time ever, I felt an implicit trust in my capacity to guide myself through the incredible labyrinth of dark and light.

It is this experience of trust that is perhaps what I value most from my journey. It has shifted my perception greatly regarding the nature of life and serves as a constant reminder of the wonder and beauty of the whole of my journey—the painful as well as the beautiful. It also served as a reminder of my capacity to navigate even the most treacherous of passages.

On a physical level, my experience felt extremely rich and layered, as if the ayahuasca was winding its way down into the deepest parts of my consciousness, as well as down into the deepest core of my body. My experiences became less clearly defined and more visceral the deeper I moved into the journey. I felt the medicine to be much like a snake, traveling from my brain downward, finally lodging in my groin. As I came to the end of my experience, I felt rooted in some tangled, steamy jungle, rich with the scent of death and rebirth, slowly becoming one with the vines and the very earth itself.

It has been five or six years since my experience with ayahuasca, and the writing of the above account. I have not taken this medicine since then and do not often consciously reflect back on its impact. And yet I know the experience is still very much with me. There were a few moments during the journey that I consider peak experiences, moments that continue to inform my choice and attitudes. The trust I felt in my own courage and wisdom was phenomenal. The joy I knew in the face of every experience, no matter how horrific or celestial, was immeasurable. I feel the ayahuasca was catalytic in my coming to know an openness to myself and my life on the very deepest of levels. There were powerful lessons in this opening. I feel now it is a matter of choice, as I consciously attempt to move toward this full embrace every day that I am alive.

Journey to
the Emerald Forest

RICHARD N.

A therapist and teacher in his fifties describes a profound physical and emotional healing—releasing trauma residues from an accident, childhood, and birth—through his encounter with the snake and the jaguar spirits mediated by ayahuasca.

I went to the circle and lay down, on my virgin voyage with ayahuasca. I was expecting some sort of an electrical jolt at the onset of the medicine. I kept waiting while my body began to warm up. I felt a gentle, subtle energy at the base of my spine, and then I felt energy in my head, more gentle and subtle yet. Meanwhile my abdomen began to warm up and move. I kept looking for the jolt and there was none. I struggled with the idea that I was not having the experience I was expecting. Then I realized that I was feeling the energy, that I had accepted the medicine, and that it was working in me. I stopped struggling.

The energy became more pervasive, moving up through my whole abdomen and back. Then I became aware of two areas where the energy was stronger than any other place—my lower abdomen and my head. My awareness bounced back and forth between the two. Then the energy in my head would become more intense and my attention would go there. I would begin to have faint visions, but immediately the energy in my abdomen would become stronger, and my attention would go there, and I would feel more intensely in my abdomen, and I could feel my abdomen opening. I could feel and hear the gurgling as I traced the medicine moving through my opening abdomen. I have had, ever since having hepatitis, mild pain in areas near my liver. I could feel the medicine going to these spots; the spots of pain dissolved and have not come back.

The energy began to move strongly up my back, and I realized I was moving with it. My whole body became involved in a slithering movement, and I was not in control of this movement. It felt good. It felt freeing. This went on for some time before I realized that I was a snake slithering across the ground. I asked the snake to show me what I was to learn. I continued to slither across the ground. I asked again and again. Again and again I continued to slither across the ground. Again I asked. This time I began to move on all fours. I could feel how what I had thought of as one part of the body moving was the whole body moving. No part could move without the whole body moving, since no part is separate from any other part.

As I moved, I felt power, the power of long sinews, of powerful muscles stretching and contracting, of four paws solidly connected to the earth, of blackness, of sleekness, and of shininess. I felt the power of the jaguar or black panther. I asked the jaguar to show me what I needed to know. Again I felt his movement, powerful, sleek, with the ability to explode into movement, leaping, running, charging, or walking solidly connected to the Earth. I asked again to be shown. I went back to being a snake moving rapidly along the surface of the Earth following every curvature and crevice. Again I asked to be shown. This time I, as the snake, raised up and I felt a hood above my head. I noticed that in sitting up I had withdrawn my arms and put them against my chest. This was what it was like to be without arms, without shoulders. (I had dislocated my shoulder in a serious accident some years before.) At this point I realized that I had disassociated from my shoulders. In doing so, I had not let the energy flow into them most of my teen and adult life. Even more since the accident, I had been withholding more energy because of the pain.

I asked the snake to show me how to heal myself. I became the jaguar again. I felt the powerful energy surging through his/my body, through his shoulders. I felt what it is like to have the whole body working together. I felt what it is like to have powerful shoulders, to have the whole body move together multiplying the power of the body. I asked the jaguar to show me again. I saw/felt the jaguar solidly connected to the Earth as he walked. I saw/felt him not as *apart from* the environment, but as *a part of* the environment. There was no separation between the jaguar and his environment. The Earth supports the jaguar and the jaguar feels himself to be the Earth—there is no separation.

After ingesting the medicine for the second time, about an hour and a half after the first, I was again aware of the energy moving through my whole abdomen and through my whole back up into my skull. The jaguar/snake dance began again, and I kept asking to be shown. I saw a peacock with his tail flared. My immediate feeling was one of dislike, contempt. I felt that all he could do was preen himself and that was not good enough. He had to do something else. Then the understanding came that it was quite enough for him to be a thing of beauty, quite enough for him to be what he was. All he needed was to be himself. I puzzled about this, feeling incomplete in this exchange. I realize now that I could have asked and felt what it was like to be a peacock, but my antipathy to simply being without doing something kept me from that experience. I missed out on peacock consciousness. This is something for me to explore, something for me to grow from. I did ask the peacock to show me what I was to learn. I became some sort of small, long, fine white-haired creature. The feeling was of tremendous luxuriousness. The feeling of this hair went all through my body, and I became luxurious.

At this point I turned onto my left side. I used my left hand to knead my right, wounded shoulder, and I was aware that my left side was trying to heal the right. I was aware of tension in my right side that I had never experienced before. My attention went to my right shoulder where my left hand was kneading. I realized that the right side distrusted the left, and this created the tension in the left and kept the left from healing the right. What came into my head was a reoccurring dream I have had since as long as I can remember. It is a dream where I am caught in a snake hole and cannot get out. In it often I get my head out; sometimes I can only see the light from the opening. In the dream I am stuck and struggling but getting no place. I understand this dream to be about my birth, being anesthetized with my mother while I was in the birth canal. It is about never completing the birth process and about feeling that I have to continuously struggle.

However, in this journey I am aware for the first time that I do not trust the feminine because I was stuck in it during birth, that I am still stuck in it. The feminine has caught me, forced me to struggle, threatens to swallow me up. I distrust all that is feminine, the mother, women, the body, the earth. Realizing this, at this moment, I somehow let go of the distrust. I feel the tension go out of the left side (the feminine side). I feel a kind of acceptance. Then I turn onto my stomach, and I feel a connection to the Earth for the first time, and I feel connected to my body in a way I have never felt before. I can let the Earth support me. I can let my body support me. I can trust both.

At this point I went to the bathroom and defecated. I have never felt all of the muscles in my abdomen and pelvis convulse in such a wonderful spasm of release. I released something and was released at the same time. It was as if the whole universe was releasing and being released while this was happening. I wanted to do it again but could not. I knew that this was it. It was astounding. I came back to the circle and laid down. I remember feeling the medicine going deep into my body. I remember its gentle penetrating warmth. I remember feeling that I was being held in its gentle rocking warmth.

We went to bed, and I lay in this warm, ever more deeply penetrating energy, feeling held all night. Occasionally I would awaken, perspiring, yet feeling totally at home.

The next morning my shoulder did not require the usual exercises to enable me to move it without pain. Neither did the usual sciatica bother me at all. In the following days my chest opened up, my shoulder increased in movement and strength. It is not healed, but it is progressing. I have been without acupuncture for two weeks, and my shoulder is better than when I go weekly.

In my day-to-day life, I am letting go of struggling. The clinic where I work has moved, and all of the therapists are becoming very territorial. I felt myself moving toward that stance too, for a moment. Then I realized I do not have to struggle about things like this. My old worries about financial survival have diminished. When I have a crisis, or what I would have seen as a crisis before, it no longer seems so critical. Instead I feel differently. Yet, I cannot say how. This is a mystery. I do feel more and more like the jaguar/panther. I feel his power, his sleekness, his sinewy strength, the energy he can unleash into explosive movement, his oneness with the environment. I feel his instinctive knowledge that the Earth and environment will support him.

What I have brought back from my journey to the Emerald Forest is the Emerald Forest, a place where all is environment, nothing is separate, where there is a mechanism that is old/rediscovered/new in me that allows me to be in peace, confidence, power, and whole.

Before another journey with ayahuasca I had several intentions: to meet with the spirit of Frog again, to further heal my shoulder and back, to be freed from whatever is holding me back. However, when I went down to the beach to meditate upon the intentions, I could not bring any of them up. Instead, spontaneously, came the words "bring me new life," or "bring me a new life." All of the time I was down there, nothing would come so I accepted this. When we stated our intentions in the circle, I could remember the original intentions, and I kept the new one in my mind.

In a previous journey, I had seen a giant frog in profile just at the very end. He had taken no notice of me. In this session, I saw the frog again, front on. He saw me, and he showed me how to use the joints of the body. He did it through swimming. He demonstrated, and then I became the frog and was moving my joints the way he did. There is such wide movement in the joints of frogs, and there is such agility.

When I was in Mexico some time later, I used the experience of the frog when I was swimming, and I found I could extend my arm fully and that I had full motion in the water. Ever since then I have felt the breaking down of scar tissue, which now is the only thing keeping my shoulder from being healed. I am learning that breaking down scar tissue is a long, slow, tedious, and painful process.

In this session, I remember being in an ecstatic state most of the time. I felt radiant with energy and transformed into a godlike being, which was totally blissful. Most of the session was spent in just being this radiant being. It was as if I was learning *to be* in a different way that was not at all conceptual. It was as if I was being re-formed, re-wired, re-made, and that in this process there is no content for the mind to work upon, only the experience itself.

At the end of the session, I became aware of a beautiful, art deco-like, brilliantly plumed being. While I was looking at it, I realized it was Quetzalcoatl. The session ended with this vision.

In the months since this experience, I have had episodes of illness, including intense fevers, sweats, weakness, depression, hopelessness, insomnia, loss of appetite, and other symptoms. I have also had respiratory difficulty. My lungs have always been my weakest organs, from childhood onward.

What has happened internally is that I have let go of numerous conscious belief systems, and I have felt freed by this. It has allowed me to act on the spur of the moment in ways that I have never been able to do before. I can see opportunities in situations that I have never been able to capitalize upon before. My life has become a constant flux within and without, and I like this.

The clinic, where I have been working, has gone bankrupt. As soon as I can responsibly release the clients from the clinic, I shall do so, and I shall be out of there. I do not have any concrete job to take its place. Yet I am looking forward to getting out of there and to whatever may come next. A year ago, I would have been in a very high state of anxiety about all of this but now I am looking forward to greater freedom, and the discovery of new opportunities that this makes room for.

Looking back twelve years later, I hardly recognize the person described in the preceding passages.

Physically, there is no pain in my shoulder and I can do every swimming stroke except the butterfly. The sciatica problem was nearly resolved until I had an accident one year ago. Psychologically, I now follow my instincts and am more spontaneous in speech and action. Professionally, I have created a private practice and am teaching yoga to the public as well as in retirement centers. I am a consultant to one of the largest HMOs, applying yoga to serious mental illness, and writing a pilot study.

Whereas before I would analyze a situation before moving into it, I now find myself in new situations without forethought. This is exciting, rewarding, and fulfilling. Daily, I am discovering who I am.

The Long, Multifaceted Journey of Jewish Experience

ABRAHAM L.

This account, by a therapist in his fifties, combines subtle phenomenological observations, psychological insight into relationship patterns, and a conscious connection with Jewish ancestors that revealed to him deeper perspectives on the story and the worldview of the Jews.

My intention in the first trip was to access directly inner teachers for my guidance in this world. In general, how to station or position myself for my purpose and specifically, any instruction for dealing with the several realities or interactions with which I'm engaged. I particularly wanted to *see* an objective manifestation of my teachers. I also wanted to experience a direct and clear verbal communication with the teachers.

As something in the visual would begin to manifest, the patterns would shift from the uniform geometrical designs toward more specific recognizable forms. Not that I would see the form exactly. I would *feel* that this was happening and trying to see, to make out the emerging whatever. I would reluctantly recognize that something in me was resisting the complete emergence. I would be tensing up, contracting, fearing, or doubting the process. I'd struggle with that process, trying to bring the vision on. Soon however, the tension would become a preoccupation (or some thought associated with the tension would grab my attention), and I would (sometimes consciously, sometimes not) shift into dealing directly with the healing of wounds and pain, and "clearing obstructions."

The entire session seemed to be a cycle of this sort, definitely a long working session.

Several times, on focusing on visions, a large serpent head would appear out of the patterns. I could bring it into greater focus, but I had difficulty bringing it closer. I wanted to be swallowed by it or swallow it, but I could feel some of my resistance to this. I got messages of the need to purify my intentions. At one point, seeing the serpent outside the window, it opened its mouth very wide, and I found my own mouth stretching wide open, my tongue stretching out, my throat opening wide, and I became aware I was being taught to open my throat center more.

I did have one experience while very tired of just wishing the serpent to swallow me, and it came forward, and I felt myself tumbling down its throat but lost focus and attention at some point in there.

At another point, I became aware of teachers working with me and the group that I could not see but feel. This helped me to relax and open more than when I was so focused on the visual. I also could then look at (feel into) my visual system itself and feel that it was very clogged up, and there were belief systems working against clear visual perception. There was some very heavy energy around the eyes and optic nerves, and it seemed to relate, at least partially, to prior programming in my spiritual training that was against "visualization." I could understand the tension of wanting to see and not wanting to visualize.

One ancestor spirit that appeared was my wife, the mother of my children, who was killed in a car crash seventeen years ago. I experienced letting go of some of the last images I held of her and relating to her as she is now; and feeling the support that has continued from her for the raising of our children. It was a peaceful and beautiful experience of love and thankfulness and a promise to be more conscious of her help and guidance.

A major lesson in the whole experience was an increased compassion for my body and the various states of consciousness or identities that I would assume. In past experiences, the intensity of not being in the "correct" or "enlightened" state would tend to cause more resistance, anxiety, or even panic. Here it seemed easier to just accept the process; after all, the processing was clearly endless.

There has been considerable carryover into my experiences since the session. I've been conscious of the continued presence, inner work, and support of the teachers and spirits contacted. I've also been conscious of a far deeper human feeling connection in my personal relating. The two aspects have been integrating and feel a part of the same lesson: to purify intention, go deeper, and bring more forth.

With the support of my wife's spirit, my family relationships have gone through some dramatic upheavals and very positive resolutions, and I have never felt more clearly connected to purpose and inner direction throughout.

My intentions in the next session were to further explore the visionary world and develop a closer relationship with the Spirit Beings. Also, to explore my ancestry as a Jew. And finally, to explore issues of addiction and its treatment (the field in which I work).

I found myself focusing on a pain in my heart. I had thoughts about past relationships. Then I started thinking about my relationship with P., my present partner. I felt it was wrong somehow, all our disagreeing and fighting. I was embarrassed and ashamed of our relationship. I saw that I wanted to blame it on her, but whoever's fault it was, it was wrong. I felt like this was the load on my heart: being with the wrong woman. I had to end this. I could never be whole, be true, be enlightened, with the "wrong" woman. I then had a vision. I saw the relationship, and I watched it go through a transformation, and then I saw it end. We came apart and stood before each other and saw that the relationship as it was now had ended.

This was a bit of a shock, and it brought me to here-now reality. I knew that P. was sitting right next to me in the circle; and I doubted she knew what had happened to me. I looked up, and she smiled, friendly, not knowing. I thought about telling her and how we'd work it all out, separating and all. Then I thought about this trip being familiar, that I'd done this before. When I checked back in, the pain was still in my heart.

Then I had another vision. I experienced myself as a knight with a sword, riding into a storm of "demons," or static dark energy forms. I was riding some creature and doing battle with the "enemies" in my heart, and at the same time, mythically, fighting for the maiden, the woman, P., my own other side. I saw the "problem of relationship," the great and difficult learning process that everyone is somehow struggling with, that has so much fear and anger around it. It felt good. At a human, personal level it felt good. I had fought for my woman. An ancient feeling.

After this experience my heart opened to P. in a way that it never could before. I experience her more now like when we first came together, only with greater understanding, and less judgmentalism. I'm able to consider our relationship growing and deepening without freaking out. We've acknowledged our inner engagement to be married together.

Along the way there were many visions of the ayahuasca serpent god. An awesome, majestic being, both within me and around me. I was also inside the serpent. At points of difficulty, tightness, and confusion, I would practice letting go and ask for the meaning of the experience. Then I would feel an opening and the presence of the serpent; sometimes a glimpse of a more extended relationship with it. At one point, riding on its back, we were moving incredibly fast through the universe, then I started to get scared and things slowed down.

There was still a pain in my heart center. I moved into it and remembered my father. He died of a heart attack. I asked myself, Do I need to grieve? My energy felt "dry." I felt things but not in a personal emotional way. I could feel the pain of unconscious feelings, and I could move into them in a healing way. I moved in and remembered my ancestry, the lineage. A long trail going back—the Jews. I saw my association with being Jewish as anguished and sorrowful, a contraction in the heart, a deep, deep pain.

I saw and felt these masses of Jews clinging to something in their hearts. Clinging to grief, like an addiction. Holding onto it as though it was something precious, something that made them special or closer to God. Perhaps the thought they had was that they are chosen to carry this burden for everyone. My heart was getting lighter as I became more objective to it all. I saw people, Jews, pressing up closer to the Wailing Wall, straining and groaning with pain. The pain being the ticket to get closer to God, the wailing being the song that carried the communication upward. A communication of the misery of the burden, that we are doing our job.

I saw through this vision to a deeper vision, as though I had moved through the wall. Here were the great mystic rabbis laughing, dancing, and singing in ecstasy and joyous celebration. Celebrating God in nature and human experience. They were not at all judgmental, not even of the Jews wailing at the wall. They were celebrating that too: they could see that it was their struggle, their journey, it had a purpose. I saw the long multifaceted journey of Jewish experience, with so many eras of suppression and bare survival. The marches through the desert. A mouth reaching forward, totally dry, my own mouth, stretching out for a drop of some liquid that might satisfy. Knowing it is not the physical water. Knowing that only one drop, if I let it touch my mouth, would wash through my whole body and quench this thirst. I would be blessed by it dropping into my mouth. Honey from the rock. I felt it come down into my mouth, and I swallowed it and felt good.

Another vision: I saw many Jews, mostly old but some young, in their yamalkes and tallises, all in rows praying. They were facing the ark where the Torah is kept, and the curtains were open and the Torah was channeling light down into them. The light was from God, and they were receiving it through their various levels of understanding and openness. It was nourishing them and keeping them alive, allowing them to survive, giving them strength. I felt that light move through me, and it felt very deep and good. It was healing and teaching. At this moment, I think I will try to acquire a Torah to place on an altar at home.

A postscript about this vision of Jewish history: I've been very moved, angered, and saddened by the current crisis in Israel, where Jews are relating with mean-hearted rigidity and cruelty toward the Palestinians. I believe what was revealed to me points the way toward an understanding of this pattern: the pain of self-imposed suffering overflowing, attacking us from within. This feeling of being attacked reinforces the defensive walls that surround the heart. In truth, there is an inner battle that Jews need to wage to become free of their present conflict. Perhaps, like with me, there is a need to fight to liberate the inner feminine. I am experiencing my racial/ethnic karmic patterns and working to process them. I'm also questioning what, if anything, I can do outwardly to help.

These experiences triggered a much deeper feeling of connectedness to my Jewish ancestry, which has opened me to a greater feeling of self-acceptance and a closer relationship with my family. I have entered into the study of Jewish spirituality and brought this into my own meditation practices and experiences in nature. I continue to feel dedication to healing the heart of "my people."

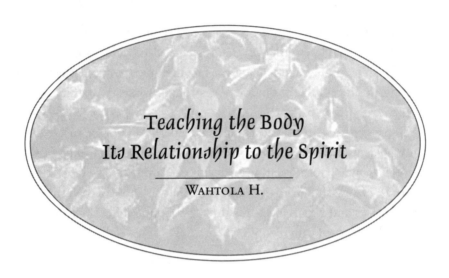

Teaching the Body
Its Relationship to the Spirit

WAHTOLA H.

This is an account by a chiropractor in his forties, who experimented both with synthetic DMT and with the ayahuasca plant potion that is the botanical parent of this molecule. He experienced the "agony and the ecstasy of creation, birth and liberation." A deeper connection with the feminine side of his lineage was followed in real time by a synchronous, unintended visit to his mother's and grandmother's graves.

This first experience took place in Mexico and was preceded by an experience of profound cosmic-orgasmic consciousness triggered by the inhalation of some DMT [the synthesized active ingredient in ayahuasca—Ed.]. My intention was to open to the feminine for healing of self and extensions of self that relate to Earth. Ayahuasca, the bitter-sweet tasting medicine went down so easily. Early in the experience, I was taunted with subtle visions of seductive creatures, women, and various scenarios. No nausea, but not enough power to fully let go. I asked the guide for a push and took another toke of DMT.

Immediately there was full penetration of the ayahuasca spirit deep into the deepest recesses of my being. As the plant teacher dug deep into the marrow of my body, I convulsed in the agony of the overpowering feminine power as it coursed through my root demanding surrender. Only after assuming a birthing position, on my hands and knees, with my face in the bowl, did the convulsions proceed to purge my deepest darkest recesses from my toes, genitals, and gut into ecstatic release of all resistance, trauma, pride, fear, all of the armor that kept me sealed within the sepulcher of my identity.

I became Kali giving birth to the Earth, through my pelvis and my mouth simultaneously. This birthing freed me from all that identified me with *maya*, purged me of all that resisted the Divine Mother. The Mother of the Universe made violent rapturous love to me. After humiliating me beyond description, she lifted me up to the barest fullest essence of self. As this occurred, simultaneously I was flying with abandon through the jungle, speeding past lush vegetation, animals, jaguars, birds, snakes, and other luminous beings on a kaleidoscopic roller coaster.

I deeply imbibed the agony and the ecstasy of creation, birth, and liberation.

In that totally open, vulnerable position of birth-giving on my hands and knees, I had fulfilled the destiny that the *I Ching* had prophesied. With the facilitation of tremendously heightened awareness, I had "bitten through" the frozen patterns that held me in prison and opened my self to the "nourishment" of the divine substance of *nirvana*.

The fragrance from the purge was nectar that wafted up out of my bowl through my open mouth and nose to warmly suffuse my being with the substance of the gods. So fragrant, sweet, and so healing was the essence of my initiation. The headless lady, Kali, taught me well, taught me deep, made me surrender my deepest recesses to her touch. Such deep bittersweet depths of release, depths of dark, moist nourishment, saturating my being, blessed sacrament of the headless lady.

Medicine Circle in California Mountains

This second experience took place in the mountains of southern California. My main intent was to establish deeper mind/body connection with the intelligence of the Earth. The journey started with subtle physical symptoms, and brief flashes of colors—bright purples, greens, and more subtle images. Image-thought patterns—of climbing cliffs, walking precarious ledges. Realizing they were visions, I got comfortable and embraced the experience. I took a booster when it was offered after an hour and a half. Very quickly the intensity of the visions heightened. The flow of the visions became constant and intense—brightly colored asymmetrical patterns, with occasional female forms, serpents, and plantlike images. My lower centers, my thighs, pelvis, and abdomen became the focus for the waves of fluidlike power that pulsed through my spaces. This fluid power was metabolizing and restructuring the consciousness of my body in relationship to the power and awareness available to it. As the intensification occurred, the purge was stimulated, without the deep significance of the first time with ayahuasca.

The guide gave clues that helped me greatly in reestablishing a center of focus when the flooding effects of the medicine would space out in directions that I didn't want to go. One set of instructions related to the four things useful to remember on a journey, inner or outer: one is your intention or purpose, two is your ancestors, three is your light or awareness, and four is the Earth. This repeatedly brought me back to center so that I could cease grasping or trying to interpret, or trying to attain a particular goal. Instead I could enter into the pure encounter with the fluxing forces that were flooding through me, informing and transforming my consciousness to the preexisting inherent identity that is fully conscious of its interconnectedness with all things.

One major effect of this session was an increase of my ability to tune perception to the life force that fully supports and provides me with experiences and opportunities exactly as needed moment by moment.

The day following the session, while I was driving home and still feeling extremely open, I spontaneously decided to take an alternate route. I was totally unfamiliar with this route but knew that the 605 would get me to the 405, which would take me home. On the drive, I suddenly became aware of a sense of familiarity in the hills I was driving past. Gradually, their significance emerged as I realized I was driving by my mother's and grandmother's gravesites. I decided to follow the inner direction as it flowed. When I stopped to get some flowers, I couldn't stop the flood of tears, flowing from such deep emotional levels of recognition. I was absolutely open to the river of opportunity that was flowing through me. I found the graves, which were close together. I hadn't been to this site since the burial of my grandmother in 1987. My mother was buried there in 1972. I had never been there alone.

I first communed with my grandmother and experienced deep levels of resonance and appreciation for her as an ancestor. I communicated my appreciation to her spirit for her contribution to our lives and for the special gifts she brought to my life, including the healing energy I could sense I inherited from her.

As I approached my mother's grave, there was an entirely different emotional tone. Deep tears of joy and longing to be with her welled up in waves, as the spirit that brought me to this experience pushed me into a deeper rapport with my mother's physical remains in her coffin. I experienced the remains of her physicality from an energetic perspective, sensing the genetic tone of her remains. I saw how her body had housed her spirit, providing her with a vehicle of expression and the means for her purpose of incarnation. Part of her purpose of incarnation was to provide me with the qualities I needed to unfold my purpose. I could sense how her qualities contributed to the interweaving with my father's and my essence to bring me to where I have come on my path.

When I tuned into her spirit essence, our lighted natures unified in a profoundly deep embrace exchanging unconditional love and acceptance. I felt such reverence for my mother as I perceived the stature of her spiritual nature surpassing that which I was aware of in our earthly relationship. In this experience, I felt that the processing that was accomplished in the deep reunion that took place affected both my mother and me, but also included all of our ancestors.

The two sessions with ayahuasca had long-term progressive effects on all levels of my spirit, body, and mind. My body continued to gain flexibility and deeper awareness particularly in the pelvis, legs, and in my relationship to the Earth. I have more continual awareness of the relationship of spirit and body in many ways—particularly in my healing work and my overall sense of interconnectedness with the ground of being. The pathways that were opened by the medicine have remained, and the processing that was done physically was assimilated very deeply. I am thankful for the opportunities with which I have been blessed.

In the ensuing two or three years since those sessions, my spiritual practice has digested the fruits of the ayahuasca experience, which I now see as actual encounters with the Great Mother. My path has incorporated the awakening of my magnetic-receptive side in very tangible ways, as the vehicle for Spirit to fully manifest in my life. This has brought forth deeper healing abilities in my work as a chiropractor. My awareness of the physical aspects of states of consciousness has evolved to the point where I can use my physical perceptions to identify how and what my relationship to Spirit is.

My spiritual quest is truly earthing in the elements of my physical reality. All praise to Father-Mother Spirit.

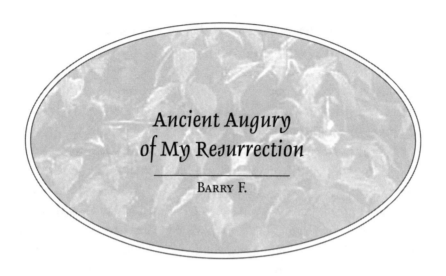

Ancient Augury
of My Resurrection

BARRY F.

A therapist in his forties relates his first ayahuasca experience, in which he passed through an almost convulsive purging that released years of accumulated psychic negativity and energy blockages, accompanied by brilliantly-colored visions and a profound affirmation of the value of earthly life.

I was very excited about the experience that I might have with ayahuasca. My friend B. had, nearly a year earlier, described his first experience with this medicine. Even at the time, as he was relating his experience, I felt waves of energy and an intuitive sense of the potential of the medicine. So, as the time of the session approached, I felt that my own journey would be a powerful one.

I entered into this quest with the intent to stay grounded in my being. In past experiences with psychoactive substances, I usually had an implicit agenda of leaving my body and this world to experience transcendence. In reality I was also attempting to distance myself from my personal pain. In the year and a half that I had been living in California at the time of this first session, I had come to realize—through my own therapy, my studies, work in the field of death and dying—that it was only *through* my experience, including the pain, that I would grow as a being. So I had already concluded that being grounded, being with the pain, facing my fears, embracing my embodied life was the work I had come to do.

The first effect I noticed was the medicine moving through my bowels. I closed my eyes and saw the spirit of the ayahuasca as a fiery red dragon flying through the inner space of my body. I heard it roar as it flew through my intestines, purifying them. I knew that sooner or later, I was going to vomit. I waited with a mixture of anticipation and dread. I knew what feeling of freedom and lightness would come after I had purged, but I was also terrified of the pain and struggle that might precede the purging. I had not purged in almost twenty years. My early recollections of vomiting as a child were all filled with unbearable pain. As a child I was frequently sick and usually vomited when ill. That purging was always an arduous struggle in which I felt that my very being was being expelled. As I awaited the purging now, I realized that those arduous struggles of the past were symbolic of death, a death that I both desired and dreaded. I desired death as a child because the emotional pain I had to endure was so intense; I wanted to vomit myself into oblivion. On the other hand, the struggle to hold on, not to die, was a manifestation of my desire to live. Caught between these two poles, vomiting as a child was always agony.

So, as I approached the time of purging, I realized that I must simply let go of my fear of the process. As I drew closer to release, desire and fear commingled in my mind until I could not tell them apart. Then it came, and I let it come. I maintained my upright posture and relaxed my body and mind as much as I could and flowed with the irresistible. It was like giving birth to the excrement of my life—clean, pristine excrement. I loved that vomit: it was my salvation, my purification, the ancient augury of my resurrection. It made everything all right. Looking upon my vomit in the beautiful bowl of my life, it was as I very much imagine the moments after death to be: looking upon the work and struggle of a lifetime with a love that redeems all experience in tears of gratitude for the life we have led, and that redeems the suffering that ushers in our becoming. So I cried at this wonderful purging and put the bowl down and continued to sit with a growing sense of the utterly simple grandeur and mystery of my life and of life.

After I had purged, and as the music of the ayahuasca *icaros* continued to play, I began to experience a radiance at my crown center, at the top of my forehead and at the brow center. This light remained throughout the rest of the journey. I watched the light of creation spiral round and round in the center of the circle. We were all like children riding the merry-go-round of creation, rejoicing, laughing with God. It was blissful. And it became clearer and clearer to me that the source of the love and joy I was experiencing was God and that God was in me, the ground of my being. So that I would share in the collective energy of the circle and act, psychically, as a midwife for the purging, purifying pain of others and then return to the source of love within myself. I felt myself embracing myself in a way that I have never experienced in my life before, as far as I can remember. I felt completely grounded in my being and filled with love.

I also had never experienced the bond of common humanity as I experienced it that night. Knowing that the pain and the purging were the common lot of all sentient beings, and of all human beings. And that we could only accept and compassionately help others to accept. I felt myself surrender to my humanity and I felt my power and my divinity there.

In subsequent years, I participated in other circles using different medicines to explore my psycho-physical being. I ingested ayahuasca only once more, in the last vision circle I attended. In that circle I received the clear message from my body that it was no longer useful for me to ingest these various medicines, that my exploration needed to proceed in other directions. Nevertheless, my experience with ayahuasca, and indeed with all the entheogenic plant substances, taught me lessons that have remained and deepened within me over the years: the fragility and sacredness of the body; the trickster quality of the shadow, and indeed, the universe, which impels us to grow despite our cleverest obfuscations; our connection to the ancestors, who keep us rooted to this Earth; our relationship to one another, in service; and a commitment to wholeness and enlightened awareness.

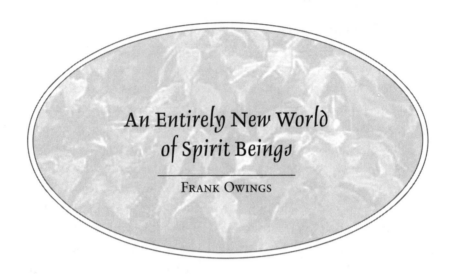

An Entirely New World
of Spirit Beings

FRANK OWINGS

This is an account by a foundation executive, in his fifties, whose love of the Pacific Ocean was reflected to him during the ayahuasca journey as a spiritual love relationship with the goddess of the ocean. There ensued a cosmic vision of the oneness and magnificence of all life.

On the preliminary vision quest I became aware of a desire to purge myself, as though there were something, perhaps from early childhood, that was holding me back and could be released. A fear, a pain, or a way of being, that I believe is related to my lower back pain, and perhaps to my very early relationship with my mother. It feels like a block or obstacle that I would do well to be rid of. I want very much to move forward in my spiritual life, and I feel as though there are impurities that I would like to dissolve. The focus of my meditation was to become an instrument of God's peace. I would like to become as pure as my mother thought I was, as pure as I know that I truly am, and we all are.

I am drawn to the West, to the Pacific Ocean. I have come to feel that the ocean is my spiritual mother and I would like to strengthen this connection. I prayed: "Magnificent Pacific Ocean, please help to cleanse me tonight, right to the base of my spine. Help me to release whatever there is in me that makes me hold back from making the total contribution that I can. Help to rid me of any and all qualities that interfere with me knowing my deepest self, and having the courage to follow my deepest knowing."

The choice to take ayahuasca seemed clear to me, and very surprising, as I had never before been even remotely attracted to ayahuasca. But its reputation as a purgative and the association with the West were clear indications that ayahuasca was the right medicine for me at this time.

That evening as we ingested the medicine, I asked my spiritual mother, Pacific Ocean, to care for me, and to help me keep the medicine down long enough to fully absorb it. After about forty-five minutes, during which time I was lying on my back with my eyes closed, I was suddenly amazed to see my dog flying around like "super dog" doing loops. When that vision faded, it was replaced by an incredible light show—dazzling, sparkling, iridescent, technicolor, exploding lights. Then the spirit of ayahuasca appeared as a dancer, definitely feminine, both seductive and unapproachable. Every time that I would approach her, she would back away from me. Yet if I backed off, she would pursue me and taunt me.

The ayahuasca spirit soon changed into the form of a giant spider/dragon. This awesome monster also would approach me if I backed off and then back away if I approached her. In due course, I decided to approach her and see just how ferocious she really was: then she began to eat me, starting with my hand. When I did not resist, she proceeded to devour me completely. Once she had eaten me, instead of me disappearing, she did. I never again saw her as a spider/dragon.

I then had the absolutely incredible experience of having iridescence poured into my eyes. It was as though I was lying on my back under a great waterfall and this iridescent water was flowing into my eyes. The experience was so overwhelming that I was afraid at first that I would drown or gag, but instead it just flowed into my eyes. Later this same iridescent liquid was poured into my heart, filling me with a joy and magnificence like I have never experienced before or since. I also realized at this time that there is a fountain that runs from the base of the spine to the top of the head, with a spigot on it that we can control, and through which this same iridescent liquid flows. I came to understand that this iridescent liquid is our natural purity.

My birth mother came to me, and as a newborn baby I attempted to suckle at her breast. As I did so, I felt a strong force push me to the floor until I felt that I would be pushed through the floor. My mother's eyes were staring at me angrily. I tried again to suckle at her breast and received the same terrifying response. I asked her why she had not been willing to let me breast-feed, and she said simply that she had not done anything wrong and that formula was perfectly fine. I asked my maternal grandfather to intercede on my behalf to talk with her to see if she would open up to him, and she refused to discuss it with him either.

I continued to see her angry eyes staring at me and at first did not know why she was angry with me. However, I soon came to realize that she was jealous of my affection for my spirit mother, Pacific Ocean. I remembered then that my birth mother had died at the age of fifty-six, just two years after I was married. I realized that at one level she was brokenhearted that I had married, as she had always wanted me to care for her. We were able to reconcile this issue with me embracing her and telling her how much I loved her and how much I appreciated all that she had done for me as my mother.

Throughout the evening I had been calling on my spirit mother, Pacific Ocean, to comfort me. She would roll me around on the beach very gently with her waves, much as a mother rolls her baby on the bed when she is changing her diapers.

At one point during the night, I was given what felt like an opportunity to participate in a sort of initiation or passage to a deeper spiritual life. For years I have frequently experienced a kind of spiritual call, when I have said that my deepest desire is for a richer spiritual life. This evening I was being taken into a sect or brotherhood that promised me a fully rich spiritual life, which I was confident could be delivered, provided I was willing to renounce all special relationships that existed in my life. That is, I was to truly understand that we are all connected in a spiritual sense, but I first had to renounce my special ties to both my children and my wife. I found myself telling the three of them that regardless of what I might publicly say in the future, and regardless of the fact that I would no longer be seeing them regularly, they were to understand that they always would remain special to me. As I was saying these words to them, I realized that I could not say one thing to them and subsequently take vows that directly contradicted what I was saying to them. At which point I realized that I was not prepared to take these vows, and that if this was the price of a deeper spiritual life, I was not willing to proceed.

During the evening I alternately felt euphoric and then nauseated. I generally had to stay lying down to avoid the nausea. I also experienced tears of joy, sentimentality, love, and gratitude. I was filled with a great sense of self-love on many occasions.

I also experienced a lot of healing around a divorce of sixteen years prior. I recognized how important my former wife's role had been in raising our children, a role that I had previously discounted.

I felt guilty that I was not vomiting and, though I felt nauseated, could not vomit. I considered going downstairs to the bathroom and sticking my finger down my throat to induce vomiting but realized that I did not want to do it that way. Finally, just as we were beginning the last round with the talking staff and everyone's attention was focused, I vomited with great gusto in front of everyone. At that point I realized that it was important for me to vomit openly in front of everyone, so as to acknowledge my vulnerability and imperfection. I realized then that the evening had been about purity on the one hand and imperfection on the other: the dichotomy of the human condition.

After the circle was closed, while I was simply sitting by myself reflecting on the evening and attempting to fully comprehend the evening's teachings, I observed that sometimes, though not always, when I observed others in conversation, I was able to see broken lines of light that went from one person to another. I was able to squint my eyes in a certain manner and see them regularly. At this time I also was observing some members of the group who were participating in a game of sorts in which they were placing one hand on their own hearts and the other on someone else's forehead, presumably in some kind of energy transfer. As I watched them, on one occasion I thought that one woman had put a decal on the other's forehead, because when she removed her hand, a rectangular patch, about one inch by two inches of luminescent light was left on the other's forehead. However, as the patch dissolved before my eyes, I realized that this experience too was part of the phenomenon that I was experiencing of being able to see energy in a way other than I normally do.

This experience remains today, several years later, the most vivid, dramatic, and real experience that I have had with spirits. The Pacific Ocean, my birth mother, my maternal grandfather, and ayahuasca were all very real teachers and participants. An entire new world of spirit beings opened to me.

In the time that has passed since this experience, I have come to believe that I was not ready for the deeper initiation that was offered. However, this was not because I really had to choose between a deeper spiritual path and my relationship with my wife and children, but because I still had too much ego attachment to those relationships. I was not ready for a deeper spiritual journey because I had so much of my identity invested in those relationships.

Several years later, it feels as though I am still learning from this experience.

Death and Rebirth in Santo Daime

Madalena Fonseca

This is an account by a Brazilian copywriter in her thirties of a powerful experience in the context of a Santo Daime ritual. A woman with a multifaceted religious background, she experienced the sacred marriage of Jesus and Mary, the masks and illusions of personality entanglements, her sacrifice by fire, and release into cosmic consciousness.

This experience took place in my hometown of Rio de Janeiro, Brazil, at a Santo Daime healing center. We were a group of four people taking the *daime* from a batch I had personally worked on, washing and drying the leaves. My purpose in taking the substance was to find the necessary strength to undergo the process of transformation that I knew was already happening. It was Easter Sunday.

After ingesting the tea, I lay down and relaxed. Soon my body was taken over by powerful physical sensations, starting with an intense sexual arousal. I had visions of flames in the pelvic area rising up through my spine in undulating movements. The perception of my body changed and took the form of a leopard. It was as if I actually was a leopard, an animal that has been present in my dreams for many years. I felt the alertness, the ease, strength, and total surrender to the power of instinct.

Then the fire rose up to my chest. I saw a beautiful woman I perceived as being Mary. She was looking at me with the most loving eyes I had ever seen, and I felt completely and unconditionally loved. My heart opened and expanded. She became a man, whom I recognized as Jesus, but the eyes were the same. Then they became two and merged into a loving embrace and sweet kiss. Soon they were making love, and somehow I was part of it, inside of them, in the same way that they were inside of me. Their lovemaking turned into the image of yin and yang, intertwining, flowing, dancing eternally together inside me. The image went down to my pelvic area, and I had an insight into the sacredness of sex.

When the fire reached my throat, I became aware of restraining cords pulling from the back of my neck. I focused my attention on the cords and followed them to their source. I saw a man looking at me with a very mean expression on his face. It was a very intense and frightening moment, because I realized I was looking the enemy right in the face. His features changed to those of other men, consecutively. Some of them I recognized as brothers, uncles, cousins, grandfathers, ex-lovers, and my own father. I had very intense and painful talks and experiences with each one of them, and all of them ended with the faces falling off like masks to reveal the face of the Devil. In those moments, I felt the presence of God and heard him say "This is all illusion!"

This was a pattern that lasted throughout the whole episode. It was a history of crimes, ambition, betrayal, violence, and pain. I experienced both sides of sadomasochistic relationships. I was victim and tyrant at the same time. Soon the experience took on a wider aspect, and I actually felt everything every criminal and every victim had ever felt at any point in time and space. It gave me a deep understanding of human pain. I cried rivers. All my resentments, shame, and guilt dissolved. I forgave humanity. I forgave myself.

Still focused on my throat, I followed another cord that took me to the ocean, where I met dolphins and mermaids. We started to swim together toward the bottom of the sea. There I saw a gigantic life-form, a purple amoebalike organism with undulating fringes. It had a hypnotizing power, and in the exact moment I sensed the danger, the amoeba opened its fringes to reveal an enormous mouth ready to swallow any living creature. I swam rapidly to safety. The creature seemed to be a most primitive life-form, a primordial mother ready to devour her children back into herself. It also felt like a vagina, a womb, symbolic of the danger of being swallowed back into unconsciousness.

The fire, by now in my head, was opening channels inside my brain. The heat became more and more intense as I saw myself being carried away by a group of people to another planet and put on an altar to be sacrificed. My body burst into flames, and I could smell and feel the pain of my burning flesh. When there was nothing left of me, I awakened to the ultimate reality, that I will call the Void, or Emptiness. There was neither God nor Devil, neither pain nor joy, but there was a sense of Omnipresence, Omnipotence, and Omniscience that was pure bliss, infinite love, and eternal being. It was a familiar sensation, a consciousness that laughed at the simplicity of the experience. I was beyond form, time, and space, back to the natural state of mind. I was also aware that I was All That Is and a part of the Whole at the same time.

When I came back to my body, I felt something moving in my stomach. I had a vision of millions of larvae cracking open their eggs. I felt a deep, joyful pain as I gave birth to myself. Gradually, I became aware of the room. I opened my eyes feeling refreshed and as full of life as a newborn.

The fact that I come from a country that is mostly Catholic, even though the influences of Candomblé, Umbanda, and Spiritism are present, probably explains the appearance of Jesus, Mary, God, and the Devil in my psyche. I thought it was interesting that the essence of the experience was about the process of life, death, and rebirth, and that it happened on Easter, the Catholic celebration of Jesus's resurrection. I was also very intrigued by the fact that Jesus and Mary were a couple, and not mother and son. That made me think that maybe the woman was another Mary, Jesus's beloved Mary Magdalene. To me, this was the beginning of a quest, a search for the contemporary Goddess, the feminine principle that embodies both spirit and flesh.

It affected my life to a point where I developed the work that I'm doing now, teaching dance as a means to bring the spirit back to the body, healing the split between spirituality and sexuality. After this experience, there's been a general sense of lightness in my life, more humor, forgiveness, acceptance, compassion, and responsibility, not only for my actions, but also for whatever happens to me. I have no doubt that work was done within the deep unconscious, and that it reflects in my ordinary life. To realize that the Emptiness is a space inside of me, always present, real, possible, and available is the greatest gift I got from the *daime,* which translated into English means "give me."

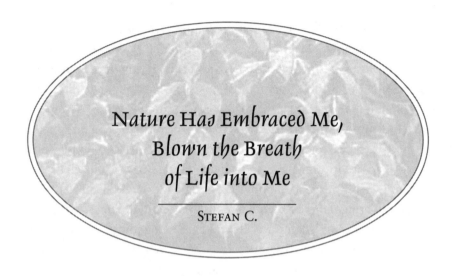

Nature Has Embraced Me, Blown the Breath of Life into Me

STEFAN C.

In this account of several experiences with ayahuasca in the context of the União do Vegetal, this physician in his forties relates various visions, both positive and negative, of relationships past, present, and future. In the course of time comes a deeper wisdom and acceptance of life's unpredictable transformations.

Shortly before leaving for an expedition to South America to participate in some sessions with the UDV *(União do Vegetal)*, I was told that a new female acquaintance, R., would be joining us. I was not pleased to hear this news. R. was recently unattached and alluring. I was encountering ambivalence about my marriage, but not to the point of wanting to leave my family. R. had apparently decided one week before our departure that she wanted to accompany me on the journey. Before informing me of this, she had purchased her ticket. The day to leave came. We traveled together to South America, to the distant Amazon. Culture shock, sleep deprivation, the excitement of our mission. Opportunities to join in ceremonies employing ayahuasca, freshly prepared from local rain forest flora. R., filled with enthusiasm, embracing our vision, bonding with our hearts.

One night, we are with a community, in a distant forest compound.

It is June 24th, the day of St. John the Baptist. Although the community practices a religion where ayahuasca is consumed ceremonially, the traditions of the powerful South American Catholic Church remain strong. For the night of St. John the Baptist, a special outdoors ayahuasca session is planned. I sit adjacent to R., in a reclining chair. I settle back and close my eyes. Visions come to me, beautiful, enchanting, soft, subtle reds, blues, and greens. My body starts to gently vibrate. In my reverie, I "see" R. and I slowly begin to elevate out of our chairs, rising, until we are suspended above the congregation, bathed in beautiful and shimmering lights of divine hue. In this celestial palace, among the union of the assembled faithful, R. and I are being joined, married. I am enraptured, overwhelmed with ecstasy. But, from the distant periphery of my consciousness, comes a sound, a question, quiet at first, then rapidly building in intensity. I listen intently and finally decipher the words. It is asking me: But what about N.? N., my wife! N., with whom I had shared the past sixteen years. N., the mother of my child. I wake from the trance. The vision of R. explodes, I suddenly fall hurtling back to earth. I am in my chair again, but stunned, desperate, terrified. My world implodes, and I am cast into the darkness, losing all orientation. Meaning loses its thread. The knowledge accrued over a lifetime vanishes, leaving only vague outlines of what had once been an identity, a life. Language loses its coherence. I know that such a phenomenon as language exists, but I have lost all capacity to use productions of sound symbolically. I know that there is such a word as "chair," but for the life of me, I have no idea what a "chair" is. Disjointed, fragmented images swirl through my nightmarish vision. I lie there, for all eternity, trapped in this whirlwind of anguish and grief. Finally, I slowly emerge, shaken. I turn to R. She too has been overwhelmed and exposed. She tells me she has "seen" all of the hurtful and cruel things she has ever done to others. She has suffered the unrelenting torment of the damned. She, too, is humbled.

Years later, as I look back on this episode, the meaning is finally clear. Ayahuasca had spoken to me, yet I could not immediately acknowledge the message. Over a longer period of time than prudence or good sense should have allowed, I failed to consistently act on the teaching and paid the price. In spite of my misguided insistence that R.

and I were fated for one another, the spirit of the vine had discerned only sorrow could grow from such a union. From its ethical core, ayahuasca spoke to me that night and revealed to me that only action based on truth and integrity can prevail. My self-delusion was laid bare, a hard yet invaluable lesson.

Some years later, I attended in North America another UDV *hoasca* ceremony. Before departing for the session, I had had a telephone conversation with a friend of mine who had participated in a session with these same ayahuasqueros several nights previous. My friend told me that he and his companions had had only a very mild experience, hardly worth the time expended. Thus forewarned of the brew's weak psychoactivity, I brought with me to the session a meditation bench, prepared to at least devote my energies to contemplative practice. I arrived at the designated house and was introduced to the gentlemen from South America. I was instantly struck by how quiet and unassuming they were. They were, I learned, from a remote rain forest town and had been devoted practitioners of ayahuasca for the past twenty years.

We gather around the table, on which lay a pitcher of dark amber liquid, emanating an acrid nauseating odor. Modest amounts of the beverage are poured into small paper cups. As we hold our breaths to avoid gagging and rapidly drain the drink, words, incantations are spoken rapidly in a language with which I am not familiar. We move to a large room equipped with comfortable chairs and cushions. I sit on my meditation bench on the floor, following my breathing, becoming focused, still. I sit. One of the South Americans, through an interpreter, tells the story of the mythological origins of his church. I begin to listen to the unraveling tale but suddenly become aware of a force field of energy that is beginning to sweep through my body and my psyche. As my being pulsates to the beat of some deep, inner force, I slide off my meditation bench and lie face up on the floor. As a field of light saturates my being, I perceive the back of my skull lifting up, allowing a current of energy, the vegetable energy of the earth, to stream through me. I dissolve and merge into the harmonious, welcoming vibrations of the plant world. I am still.

Later, I return, gradually stirring from the enchantments of nature. The particles of my being are one with the flowers, the trees, the elements. I feel purged of all burdens, of all sorrow. Nature has embraced me, revived me, blown the breath of life into me once again. I have been reborn, renewed. Looking back, I still shake my head in wonder at this initiation into the natural world. Our true identities as beings of light, at one with the forces of nature, had been revealed to me. An identity I had hardly dreamed of had been made manifest. Much time has passed, yet I remain in awe of the power revealed to me that night by the messenger spirits of ayahuasca.

On one of my travels to Europe, I had another occasion to be with representatives of the UDV. In Spain, near Barcelona, in a large, low-ceiling room, I attended a gathering. It was not an easy session for me. The setting, the disparate backgrounds of those in attendance, the hot, claustrophobic environment, all conspired to create a state of relent-less physical and emotional constriction. Well into the session, I made my way to the bathroom and tried to alleviate the mounting internal tension through vomiting. I retched but could not trigger the emetic reflex, failing in my gut wrenching efforts. I returned to my seat, sitting with eyes closed, exhausted, the energy drained out of me, waiting for the ordeal to run its course. As I slipped into a half-waking reverie, my thoughts took me to the turbulent and tragic history of the country I was then visiting. Suddenly, I sat bolt upright, a sharp pain piercing my back. My eyes still closed, I entered a vision, attired as a sixteenth-century Spanish Jew, outside the walls of a city, facing an angry prelate surrounded by armed men. I am told I have to relinquish my faith and embrace the true Church, or else face the consequences. I refuse, defi-antly. I turn to walk away, and as I do, tensing as I know what is to come, I feel the sudden, stabbing pain to the left of my thoracic spine of the lance hurled deep into my back. I collapse, pain searing through me, until the breath fades and departs, and I submerge into the sea of ancestral memory. Other personas, remembrances, glide through me, themes of strength, perseverance, duty. I slowly return to the present, silent and reflective.

A year after I had last taken ayahuasca, I received a call from an old friend. He was excited, telling me that a colleague of ours from South America was planning to visit the North. He would be bringing with him, so I was told, a tea of legendary potency, prepared in the heart of the Amazon rain forest.

These friends met me at the airport, and we drove into the mountains. When we arrive, I am surprised by the number of parked cars. I had imagined a small intimate session with a few close friends. Instead, a steady stream of younger, eager initiates. We rest, converse, wait for several hours. The session is called to order. I realize I don't know who many of these people are, I don't know their stories. I feel odd, disconnected, not quite prepared for what is to come.

We are outdoors, sitting in a very large circle around an altar. There are twenty-three of us, by my count. I am sitting directly across from the leader. He is sitting on the North node of the circle. I am anchoring the South. It is getting dark. The wind starts to pick up. We are handed plastic cups filled with ayahuasca, and I breathe in its pungent, familiar essence. I gag. Oh, I know you, rank libation of the Gods. I brace myself. I see others put the cups to their lips and drink. This is it! I look inside. Am I ready? I inch closer to the edge of the existential cliff. I look and see nothing but darkness. I am ready! I begin to drink, holding my breath, chugging the ghastly brew. I do not pause. As the fetid liquid courses down my throat, as I restrain an involuntary shudder, my mind focuses on the intent of this journey, the words suddenly crystallizing in the moment. May the experience that is to come give me courage, may it give me strength. May I find forgiveness. We sit down.

I sit for a long time. I breathe in the cool mountain air of dusk. I can feel the tea inside of me, percolating, working its way. I sit with the periodic waves of nausea. I do not respond. I know this place well. This is the time of sitting and waiting. Around me I can make out through the gloom thin figures rising from their places, staggering into the woods. I hear their bellowing, harrowing sounds of emetic catharsis. I breathe, deeply. My pulse quickens. I do not move. I wait. Time passes. I feel ill, filled with a heavy, tense torpor. I look at my watch. Exactly one hour has passed. It is time. I slowly rise to my feet. My head spins, my body sways. I very cautiously begin to move, putting one foot in front of the other, traversing with great care down the side of the hill. I walk some distance away from the circle, into the trees. I halt, stand, without moving. I collect myself, bend over, and put my hands on my knees. I lean forward, and with one, two, three efforts I heave and my stomach contents erupt and spew from their turbulent resting place. Release! I step back, breathing deeply. I have propelled that dark, foul, stifling mass from the core of my being. I look up to the night sky, filled with stars. I am lighter, balanced, free. I turn and silently, confidently walk back to the circle.

Night has fallen. It is cold. Colder than I had expected. There appears to be little ceremony, as far as I can see. We are, each of us, alone in the dark. I am cold. I have neither coat nor blanket. The wind cuts through my sweatshirt. I shiver. I do not feel connected to the group. I cannot ask for help. I sit, my shoulders hunched, chin against my chest, steeling myself against the elements, wondering how am I going to get through this night. I dig in and wait. Slowly, barely audible, a song starts to build within me. I feel a rhythmic gentle vibration, rising, and then softly uttered, intoned. A simple beat, a subtle melody, voiced from my soul. It is my song. My song! Empowering me, ennobling me, protecting me. My simple song, barely perceptible, weaving a blanket of energy around me. The cold fades, leaving a protective coat of warmth and reassurance. Throughout the night, I continue to sing my song. Quietly, to myself. Within my song, I perceive my Self, my reservoirs of power. Guiding, grounding, remembering.

With my eyes closed, I see incredible explosions of brilliant color. Dazzling landscapes. A glimpse of eternity. The beauty is overwhelming. To the side, something else. An other worldly presence. I have been in this place before. Something not quite human, but what? Plant? Insectoid? Strange. Cannot look directly at it. Cannot get too close. The Gaian overmind. I look away and move on.

In deep space. Inner space. Kaleidoscopic, multicolored images dance before me, through me. I see my face, my internal image, melt, flow, re-form, congeal. Repeatedly, a new face, a new persona, consolidates, steps forward, briefly, then fading, one followed by the next. Who am I? Which of these are me? None of them, all of them. I breathe deeply, surrendering to the cosmic wave, dancing with the stuff of Creation.

Images of R. come to me. I tense, my heart hardening. I remember my intention and see a path opening up, an avenue, of forgiveness and compassion. A history passes before me. I know I need to move on, yet I understand to do so I must come to terms with what has happened. I must learn to accept. I know this is the only way to let go, and I know it is time to move on. My heart opens. I forgive R. for being who she is. Then, my wife N. comes before me. Twenty-one years together. Twenty-one years of accumulated grievance, so much of it petty. I open to N. A tremendous wave of appreciation, jubilation for that person who has shared so much with me. I forgive N., for not being who I thought I wanted her to be. Pictures of N. and our child dance before me. Images of tender, shared moments, some in the past, some which have not yet come. I see N.'s pain, her fears, her frailty. My song, quietly, gently, percussing. Through my song I open up a channel of healing. I see this path, inlaid with brilliantly colored stones, at the end of which is N., the healing energies filtering through her, filling her with light, renewal.

The long night wears on. There are fewer people sitting around the circle. Some have drifted off. Some lie nearby on the ground. We mostly sit in silence. Occasionally small groups form, quietly chatting. I am accustomed to taking ayahuasca in tightly structured rituals. This is something else. The circle feels fragmented, frayed. Next to me, on my left, my friend T. leans over and whispers to me that she has had enough. She puts her arm around me. We embrace, merge, two fields of light energy, briefly blending, then separating. She gets up and walks up the hill toward the house. People get up and move toward the house. The circle is slowly coming apart, seemingly controlled by its own autonomous mechanism. No closing ceremony. No thanking the spirits. One by one, in pairs, drifting off into the night. I sit. I am conscious of where I am sitting. The others may go, but I have a responsibility. I am anchoring the South. I am holding the circle. The others don't seem to know, but I do, so I have a duty. My back hurts, my legs are sore. But I have no choice. I sit. There is no one near me. Only the leader remains, on the North node of the circle, and one or two others sitting next to him. I am prepared to sit through the night, if I have to. I wonder how long until dawn. I quietly sing my song, a song of empowerment, of strength. I am planted in the ground, taking root, becoming one with nature. Time passes. Finally, the leader gets up and walks over to me. He tells me that he is going up to the house for a cup of tea, and that I can either join him or stay where I am. I say I'm ready to go. I stand up and walk toward the house. The circle is closed.

I sit on the porch for a long time, reflecting on how I had come to be in such a place. Feelings of gratitude and attainment pass through me. My encounters with ayahuasca had brought me to many stations along the way. Some joyous, and some full of dread. But each had taught me a valuable lesson. Where I will travel next on the path, I do not know. I will wait, and I will be attentive, and keep my senses tuned. Waiting, for the call.

The Buddha, the Christ, and the Queen of the Jungle

GANESHA

In this account, an actor and body therapist in his forties relates the visions he had in a Santo Daime ritual, in which esoteric Christian figures, Buddhist concepts, and the goddess of the forest mingled.

The setting was the Amazonian rain forest outside of Manaus, Brazil. I had experienced ayahuasca a number of times, but this was the first time as a guest of the Santo Daime, one of the main churches in Brazil that uses ayahuasca, or *daime* as they call it, as their main sacrament. We arrived at the Santo Daime encampment just as the sun was beginning to set and were taken to their main church, a covered terrace in the middle of the forest. Surrounding the center pole of the structure was an altar with Christian icons, mainly of Mother Mary, as well as crystals and candles. Two Brazilian women had told me of the importance of the Mother Mary in the Santo Daime, that the daime visions were given by the leaf of the ayahuasca mixture, which they called *Rainha*, the Queen. Seated around the altar were the inner circle of *fardados*, the priests of the Santo Daime church, and then in circular rows around them were other Brazilian participants and North American guests. Men and women are seated in semicircular rows on opposite sides of the terrace.

237

Once we were welcomed by the *mestre*, the spiritual elder who would lead the session, we went in two lines, the men to one window behind the altar and the women to the other, where we were given our first dose of the tea, a rather pungent amber liquid. Once everyone was seated, the songs began, Christian-based hymns in Portuguese with a driving rhythm, sung with abandon by the fardados and accompanied by several guitars. The songs were constant, one after the other, and went on for several hours. I began to wonder if there would ever be silence, a chance to close my eyes and focus inward on my own experience. I didn't purge this first dose of ayahuasca, and as a result, it went through my entire system, whereupon I went to the bathroom and eliminated. Later, we were offered a second drink, which I took. I later purged this dose over the side of the railing into the jungle earth.

Early in the session, much of my focus was going into the well-being of the twelve-year-old son of a friend, who was attending his first ayahuasca session. He was doing just fine and starting to have his own visions. The songs were strong and good, generating a powerful energy. Finally the singing stopped, the hitherto bright lights were turned off, and only the candles remained lit. The contrast from the sound to the quietness sent me into a silence like none I had ever experienced. It was as if the songs had charged the air and built an energy, so that when they ceased, there was tangible etheric substance all around. I became aware of my breath. The subtle sounds of the jungle emerged, the sound of the life of the jungle. I heard the breath of those around me and together with the jungle, we all began to breathe as one Being. I experienced being the Buddha, all of nature being the Buddha. The Pali Buddhist term *paticca samupada* came into my mind, the circle of the interconnectedness of all life. All things were spontaneously self-arising, a pulsating membrane of one Life. I was in an experience of nonduality, Spirit and Nature were bridged, were One.

At a certain point, the daime surged into my third eye. It began to churn and pulsate. I saw a brilliant green mandala, the green of the vibrant jungle. It stayed there for several minutes, growing brighter and brighter.

The main vision of the session came later. I experienced a large chalice holding the entire group, a receptacle being filled with light from above. I sensed my body as a smaller chalice, being infused with Spirit. And I experienced all of Amazonia as a gigantic chalice, a cauldron of life infused with the power of Heaven.

Later, as we danced a ritualistic dance, the *bailado*, in a circular pattern around the central altar, I experienced this flower-entwined pole as the *axis mundi*, the tree of life. The worshipers were joyous, celebrating life. I felt like we were all blessing each other. Light was pouring from our eyes and hands. I felt open-heartedness and unconditional love with those present as though the Christ consciousness was coming through me, that I was being an emissary of beings such as Yeshua, Zarathustra, and Melchizedek.

As the session neared its close, and I stood in the jungle under the stars, I had the strong sense of truly standing between heaven and Earth. I felt a closeness to the stars, and a closeness to Gaia. I sensed a strong presence of Mother Mary, not in the traditional sense, but as the feminine principle that was the creative abundance of the Amazon, that was the vision-giving leaf of the daime. I felt caressed. The Rainha of the Jungle was blessing me.

Agony and Rapture with Santo Daime

RAOUL ADAMSON

In this account of two sessions with the Santo Daime group, the psychologist in his fifties, experiences a rapturous feeling of empathic bondedness with the community, the forest, the continent, the Earth and beyond; in another experience, the rhythmic singing and dancing carry him through some agonizing past-life hells.

I had had several experiences with ayahuasca in small group shamanic healing cermonies and also had participated twice in the rituals of the UDV *(União do Vegetal)*, when the opportunity arose to participate in a Santo Daime session in Manaus, on the occasion of the ITA (International Transpersonal Association) conference. The UDV cermonies are reminiscent of a Protestant church service, with the people sitting quietly in straight-back chairs, in brightly lit rooms, listening to sermons and some singing by the *mestres* sitting at a table in the center. The Santo Daime cermonies on the other hand are more like Pentecostal or Gospel services, with loud driving, rhythmic singing, and swaying dances, as well as quiet periods of prayer and contemplation.

A group of about thirty or forty Americans and Europeans were taken by bus to the Santo Daime center outside of Manaus in the forest, where about fifty Brazilian *daimistas* had already assembled. We were instructed beforehand to wear white shirts (the men) and white blouses and skirts (the women); no red or black clothing. Although most of my prior experiences had been in small groups, I had already gone through a UDV experience with several hundred people, with no ill effects, so I was calm and filled with positive expectations. Besides, I had met one of the leaders of the Santo Daime church, a man with a long grey beard, and very large, gentle, expressive eyes—and I had a good feeling of trust toward him as leader of the ceremony.

The temple or church building was a hexagonal roofed terrace, open at the sides, in a clearing in the forest. Chairs were set up in semi-circular rows around a central altar table, decorated with Christian icons. The central wooden pillar was garlanded with vines and flowers, like a Maypole. About a hundred or more people crowded in the space, generating a lot of body heat in addition to the humid warmth of the tropical night. My twelve-year-old son was with me—it was to be his initiation into ayahuasca or any visionary plant medicines. There were other Brazilian children in the ceremony as well—some as young as seven or eight. One American couple brought their infant with them, who was taken care of nearby, so the mother could nurse him from time to time during the night. Long-term Santo Daime members in Brazil regularly take ayahuasca while breast-feeding, some even while delivering their babies.

The separation of women from men in the ceremony did not bother me, as it did some of the Americans and Europeans, who were used to doing everything in close proximity with their spouses or partners. The church people apparently feel that because the experience can be so intimate and revealing, couples might be tempted to cling and huddle together, and this would detract from the group unity and synergy. For the same reason, I also did not mind the group monitors reminding one to come back inside after one has gone outside to relieve oneself, purge, or breathe fresh air. At any one time there were maybe ten to fifteen people (out of a hundred) outside; if more had gone and stayed for longer periods, the group energy field would definitely be affected.

In this particular session the format was focused on healing, so people mostly sat in their chairs and sang, occasionally rising to sing while standing. At the end there was about an hour of dancing. In other sessions, the group dances the whole time, except for an hour-long silent meditation around midnight. After everyone had lined up and received their first dose of the very bitter tea from the dispenser at a sort of counter, we sat down and the singing began.

A group of young women from Mapia, the main Santo Daime community, formed the main choir—they always started the songs, which followed a definite sequence in a hymn book, called *hinario*. Since I didn't understand any Portuguese, I didn't pay any attention to the content, and just sort of hummed along with the music, which was simple, melodious, and very rhythmic—having a similar kind of rapid rhythmic structure as the shamanic *icaros*. Some men around the central table were accompanying the women on guitars, and many of the men also had metallic rattles—so the volume was quite impressive and even overpowering at times. After a while I learned to let it carry me along.

The first "effect" I noticed was that with my eyes closed I would see the familiar tryptamine geometric visuals, but with eyes open I saw only the singers, the building, and the forest outside. Then the heat began to rise to my head, and I felt I needed some fresh air. I made my way outside, winding through the crowded seats, and once outside, just before fainting, I sort of gently eased myself down onto the muddy ground, gratefully taking in the cooler night air. One of the monitors leaned over me and kindly asked me whether I was where I wanted to be or whether I needed anything. I assured him I was just fine and just needed to rest a few moments. After about ten minutes I returned to the circles and the singing.

By now I felt the energy of the ayahuasca had sort of settled into my systems, and I was happily humming along with the songs, and rocking in my chair with the rhythms. My heart was opening more and more, and I felt empathically connected to this group of mostly strangers, singing in a foreign tongue words I did not understand. It reminded me of group sessions with the empathogenic substance Ecstasy (MDMA), in which one often feels a similar emotional bonding.

With the lilting rhythmic singing, this empathy, this heart-centered caring and joyous feeling of mutuality and connectedness, embraced the whole circle and the forested lands outside; then it expanded to all the people at the conference and the whole city of Manaus. As the verses of the song continued, the rising and falling of the rhythmic phrases generated a wavelike force-field, in which we were all embraced, that seemed to swell and recede, swell and recede, expanding over the whole Amazon region, all of Brazil, the whole Earth with its vast oceans and continents, still expanding out to the Moon, the Sun, the planets, and stars, then the whole galaxy and beyond—and at the end of the song I landed back down in this body, in this place, at this time, with these people. It was an ecstatic state of rapture—of being lifted up and out of myself—or rather *with* myself, and others: human, animal, plant, and land, into a larger community of beings, angelic, divine, spiritual, and cosmic. And then to be brought back so precisely, lovingly, knowingly, caringly, at the end of the song. And then another song would take off, generating another rapture.

In some of the quieter periods in between, I was able to take some of the enormous amount of energy that was palpably generated in the room, and direct it for healing some chronic discomfort in my head and face. I was able to trace this pattern of contraction to a surgery with ether anaesthesia that I had as an eleven-year-old boy. The aware energy of the daime seemed to open up this contracted space in my body, when I chose to direct it in that way.

The session went on for four or five hours. Some took booster doses of the daime tea. Then came two hours of dancing. We were shown the simple steps—two to the left, two to the right. The monitors ensure that one stays in the line and moves in harmony with the others. That way the movements became effortless, one is carried by a wave, like a group of seals bobbing in the surf; if people move counter to the group, the harmony is lost and it becomes stressful. As it was, I felt I could have gone on for hours.

Riding the bus back to the hotel in Manaus, I asked my son what his session was like. He said it was "cool." He did not purge or have any significant discomfort, less than I in fact. I asked him if he had any visions, and he said a giant fish, similar to one we had seen in an aquarium earlier in the day, had descended from the sky into his space. He was impressed and pleased with this visitation.

After we got back we slept for the four remaining hours of the night and awoke refreshed with the energy of a normal day.

My second encounter with the Santo Daime took place in a large house by the water, in northern California. A group of Brazilian church leaders, with singers and musicians, were traveling and conducting what they call "works," in various parts of the country, assisted by North American church members. About sixty or seventy people had assembled, all wearing white, some relatively experienced, some new to the ceremonial format.

We were given an intensive introduction, in English, to the process, with some of the things to expect and written guidelines for preparation. "The Daime offers an opportunity to align with the Divine. . . . It opens your consciousness, giving you the opportunity to experience love and truth at depths you may never have imagined. . . . The Daime opens you to what is highest and lowest in yourself. The purpose is to use the highest in you to transform the low." Suggestions for the work included deep breathing, surrendering to the energy, sitting or standing with body straight, sitting with the legs uncrossed, and feeling the ground beneath your feet. We were also told, "if you can't dance—sit, if you can't sit—lie down." There were mats along the side for people who needed to lie down for painful parts of the process.

For myself, I was used to dealing with intense ayahuasca experiences by lying down, with short periods of sitting. To sit for long periods, much less dance, was a challenge, and I wanted to learn to accommodate the heightened energy and consciousness in this more dynamic way. I could appreciate how, with large groups like this, the structure had to be well-formed, for one is in a state of heightened suggestibility and too much chaotic moving around would be very distressing and distracting.

After the initial prayers and invocations, two lines formed to take the tea from the ceremonial leader; as soon as everyone had drunk and taken their place in the circles, the singing and dancing began. Those who knew or were learning the hymns had the little booklets and were following along. I was content just to concentrate on the sounds, the movements, and the feelings in my body. The hymns were all said to have been channeled by the Founder or later church leaders during ayahuasca sessions. They are simple songs calling the names of figures such as Jesus and Mary, St. John and other figures from the Bible, as well as the Sun, Moon, and stars, and certain deities unique to this movement.

About half an hour after ingesting I noticed intense energy changes in my body, and visual patterns in the field around me, even with my eyes open. The pulsating energy of the songs and the dancing, swaying bodies filled the room. I began to feel anxious and worried that I might fall down. I went outside, where some others were purging. I did not feel the need to purge, but I felt faint. One of the assistants, or guardians, asked what I needed, and I requested a place to lie down. He provided me with a mat that was placed outside on the deck right outside the window of the room where the dancers and singers were.

I collapsed and closed my eyes and breathed deeply into my solar plexus. My visual field was filled with blood and gore, bodies contorted and broken, flesh pierced and cut with metal. I seemed to be in a battlefield of some kind, a scene of carnage. The thought flashed through my mind that this was another life in which I died in battle. There was a strange quality of emotional detachment—in spite of the gruesomely graphic hallucinations, I did not feel fear or pain. My body was shaking and vibrating, and I felt very much alive and power-full, even while traversing this hell. All this time I was very much aware of the singing dancers, in the room behind the wall I was lying next to. I could hear them and could sense the pounding rhythm of the feet of the dancers.

After a while I pulled myself up to a sitting position. One of the assistants came and encouraged me to breathe more and used hand gestures to seemingly pull some etheric obstruction out of my solar plexus. I started to hum and chant along with songs, and rock back and forth with the rhythm. I felt I was riding a wild wave of energy, or some powerful animal force, having just traversed a nightmarish killing field. I was glad to have had the singing dancers close by while going through the vision of bloody carnage. It would have been much more traumatic without that. The songs and rhythmic beats keep consciousness moving through whatever it is we're in.

I continued to sit outside, connected to the singers and dancers, singing and rocking along. My mood was serene and joyous, my body felt light and clear. Various other people also came outside from time to time, to lean over the railing to purge, to breathe fresh air; occasionally one would hear moaning and groaning, but also laughter and quiet conversation. The songs and dances continued, relentlessly. After a set of hymns lasting about half an hour, there would be a brief pause, where the names of spirits, guides, and ancestors were invoked and thanked, followed by vigorous shouts of *"Viva, Viva."*

Around 11:00 PM, the singing and dancing stopped, and there was a forty-five-minute period of total silence. People came and sat outside in the darkness, looking out over the water, watching the Moon and stars and their reflections, quietly meditating, absorbed in the stillness filled with meaning. There followed some more songs for about an hour, and then the closing prayers.

In reflecting on these experiences I saw that the songs and dances are the essence of the Santo Daime. There are no other doctrines or teachings in this church, other than what is in the hymns. Like the *icaros* in the shamanic healing rituals, which have a similar sing-song rythmic quality, they allow you to keep moving through visionary experiences, whether hellish or heavenly, and not get stuck in one or distracted by the other. The form of the concentric semicircles, whether sitting or dancing, provides a container in which individuals can go through extreme experiences of purging and transformation, and still feel supported and connected.

Some people, including some North Americans, prefer the more rigid structure of the UDV ceremonies, where the priests sing and preach and answer questions; the church members only sit and listen. For myself, the Santo Daime process seems more participatory, more inclusive. *Daime* literally means "give me," which to some may sound like a demand or a plea. In my visions I understood it to mean rather "you are giving me"—an affirmation of receptivity to the gifts of the Divine.

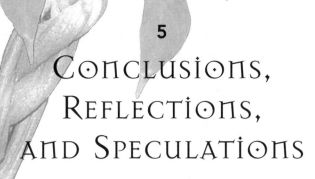

5

CONCLUSIONS,
REFLECTIONS,
AND SPECULATIONS

RALPH METZNER, PH.D.

The introduction of shamanic practices and knowledge associated with a powerful rain forest hallucinogen into Western culture in our time raises profound and challenging questions. I would like to address, if not answer, at least three of them: (1) what are the most valuable and useful applications of ayahuasca in the context of Western medicine and psychology? (2) what is the worldview or cosmology that is revealed by the shamanic ayahuasca visions, and how does it differ from the modern Western worldview? and (3) what is the significance of the resurgence of entheogenic shamanism at this particular time in the history of Western civilization?

MEDICAL AND PSYCHOLOGICAL
APPLICATIONS OF AYAHUASCA

In the context of Amazonian traditional healing practice, the drinking of ayahuasca is something like a master cure for all illness. Not that the medicine itself is a panacea, but that it functions as a guide or teacher for the human healer, pointing him or her to other herbs that might be needed, allowing him to beat back sorcerous attacks or extract poisonous infections or infestations. These kinds of practices presume a completely different understanding of illness and medicine than what we are accustomed to in the West.

But even from the point of view of Western medicine and psychotherapy it is clear from the literature and from the stories recounted in this volume, that remarkable physical healings and resolutions of psychological difficulties can occur with this medicine. Early in the twentieth century an extract of the vine was used successfully in the treatment of Parkinson's Disease, a possible application that has not to date been followed up. There have been ancedotal accounts of the complete remission of some cancers after one or two sessions with ayahuasca. Since these occurred with ayahuasca in the context of traditional healing ceremonies, it is impossible to separate out the pharmacological effect from the psychosocial and shamanic elements. Further study of such cases and cures is surely warranted.

On the psychological level also, there is intriguing evidence of positive therapeutic changes being induced by the ritualistic ingestion of ayahuasca. The research by Grob, McKenna, Callaway, and their associates with the Brazilian hoasca church known as UDV showed that there were significant differences on several personality trait measures between the long-term users of hoasca and a nonusing control group. Psychiatric interviews also confirmed these differences in that the subjects reported making positive changes in their behavior (less drinking and drug use, more responsibility and confidence) as a result of their participation in the hoasca ceremonies. As the researchers emphasize, this was not a before-and-after evaluation, following the subjects over time, so the findings are not definitive, but suggestive. The differences could also be due primarily to the psychosocial effect of belonging to the church community, with its rules and discipline.

The most unusual and unexpected of their findings however is the difference in neuropsychological functions. The hoasca users performed better than controls on short-term verbal learning tasks—capabilities that usually decline with age. This difference is unlikely to be a psychosocial effect and may mean that ayahuasca falls in the category of substances now referred to as "cognition enhancers" or "nootropics."

Certainly many of the stories recounted in this book and elsewhere in the literature support the notion that under the influence of ayahuasca people are able to see and understand themselves better, to think more clearly about their relationships, the nature of the cosmos, and their own place in it.

From the stories related in this volume, one cannot help but be impressed by the remarkable health-enhancing effects attributed to the purging action of the vine. People describe the liberating, lightening, color-enhancing, strengthening after-effect of *la purga* in near-rapturous tones. The purging gives people the feeling and bodily experience of strength, called *mariri* by some of the ayahuasqueros; and this is not muscular, weight-lifting strength so much as a kind of intestinal fortitude, a relaxed, warm feeling of being at ease in the deep instinctual roots of one's physical nature. Many first-time ayahuasca users have to overcome an initial inhibition to vomiting, because of its usual associations as a symptom of sickness. Once this is done, they find that the purging is easy and effortless and not at all accompanied by nausea or queasiness. Only where there is the presence of toxicity in the body, can the vomiting discharge become debilitatingly intense. This is also one of the reasons the ayahuasqueros emphasize being on a diet low on spices, sugars, and fats. The two most violent purging reactions I have personally observed were with a man who was a heavy smoker, and a woman who had just ended a two-week course of antibiotics.

There is an interesting convergence that often happens between physical purging and psychic purging—what seems to be a kind of discharge of negatively toned psychic contents. People who do not have any appreciable physical toxicity in their system may yet find themselves throwing up and thereby releasing the toxic residues of past emotional entanglements, the guilt and shame loads of traumatic abuse, or the self-limiting, self-defeating thought-patterns of addictions, compulsions, and other neurotic behaviors. Sometimes people might even find that what they are discharging through the vomiting is not so much their personal "stuff," but some portion of the collective consciousness-bands of humanity. I recall an individual who reported that at first he was exploring the flow of personal visions, without purging, but then he came to think about and have visions of the genocidal wars and oppression in Central America (that he had no personal experience of); he abruptly threw up.

This combination of physical and psychic purging that occurs quite regularly with ayahuasca leads me to suggest that potentially the most useful application of this medicine in Western society may be in the treatment of addiction and alcoholism. The Brazilian hoasca project with long-term members of the UDV reported a marked decline in alcoholism and drug addiction among church members—although the drug effect can't be completely separated from the effect of belonging to a structured community. Similarly, among members of the peyote-using Native American Church in the United States, it has regularly been reported there is a significant decline in the alcoholism that is otherwise so devastating to the Native American population. Here too, the return to a traditional way of life associated with participation in the NAC rituals may be equally as important as the plant hallucinogen. Looking back at the history of Western research with psychedelic drugs, the most widespread therapeutic application of LSD was found in the treatment of alcoholism. At one point during the late 1960s, there were about five or six hospitals in North America with an LSD alcoholism treatment program; the efficacy rate was, on average, about comparable to other forms of treatment.

Since the psychedelic (entheogenic, hallucinogenic) drugs and plants are expanding consciousness, heightening awareness, and providing self-insight, they are the logical and natural antidotes to the consciousness-contracting, fixating, narcotizing effect of the addictive drugs. And because of the purging effect (found in peyote and ayahuasca), there is reason to believe that such combined emetic-hallucinogens may be even more effective in treating alcoholism and addiction than LSD. The addict needs to purge, not only the toxic residues of alcohol and other drugs from their system, but also the mental, emotional, and perceptual reaction-patterns and habits. One case of self-treatment of a fifteen-year heroin addiction with ayahuasca has been reported: the woman locked herself in a room and took ayahuasca every day for two weeks, purging constantly, until she was free. The Takiwasi program initiated by Dr. Jacques Mabit in Peru treats cocaine addicts in a residential setting involving counseling, ayahuasca sessions, and physical labor in the garden. Ideally, there should always be some kind of community group for recovering addicts and alcoholics, to provide similar ongoing support after the intensive treatment phase as is provided by the UDV and the NAC.

I believe there is a strong probability that an alcoholism and addiction treatment program using ayahuasca in the context of a holistic approach that also uses nutrition, physical labor, exercise, and psychospiritual practices can be established some time in the next ten years; if not in the United States, then perhaps in Mexico or Canada, where antidrug political hysteria is less intense.

SHAMANIC ENTHEOGENIC COSMOLOGY

If we inquire into the basic model of reality or cosmology that is revealed and implied by the visions and experiences of Westerners with ayahuasca, we find that it is similar to that shared by indigenous shamanistic cultures worldwide. Particularly those individuals who have continued to use ayahuasca or other plant entheogens in a more or less systematic manner for self-healing, consciousness exploration, and spiritual practice have developed (often implicitly) a worldview radically different from the prevailing Western paradigm of scientific modernism.

Those who are ideologically committed to the still-prevailing Newtonian-Cartesian paradigm will at best consider the statements and descriptions of the *ayahuasqueros* as drug-induced "hallucinations," incapable of being scientifically evaluated or verified. From the perspective of a Jamesian radical empiricism however, the phenomenological descriptions of consciousness explorers must be accorded the same reality status as observations through a microscope or telescope. They lie outside the normally perceptible realm of sense experience, yet nevertheless they are subject to verification, comparison, testing, and replication by anyone else who chooses to avail themselves of these perception-enhancing tools. In other words, contrary to the assumption of materialist science, it is possible to be objective to one's subjective experience. In fact, Buddhist mindfulness meditation and Gurdjieff's self-remembering are practices designed to help one learn to do just that.

Those who have embarked on a serious psychospiritual practice of consciousness exploration using shamanic and yogic technologies, who are willing to trust their own experience more than the received views and concepts they have taken on faith, tend to find themselves gradually awakening to a vastly expanded and different worldview. I should point out however that many features of the traditional and newly revived shamanic-animistic worldview appear to be quite compatible with the most recent, growing edge theories of postmodern science. There is not the space here to enter into a discussion of these convergences in any detail. I will merely mention the particular relevance of ecology and systems theory, the Gaia theory of James Lovelock and Lynn Margulis, Rupert Sheldrake's theories of morphogenesis, David Bohm's "holomovement" interpretation of quantum theory, chaos theory, and nonlinear dynamics, Edward Wilson's biophilia hypothesis, and the evolutionary cosmology articulated by Brian Swimme and others. As one specific example, we can recall the evidence put forward persuasively by Jeremy Narby that ayahuasca shamans shift their consciousness to make observations at the molecular level of reality, giving a subjective view, as it were, of the structures and functions of the DNA molecular code of life.

I shall simply present here in brief propositions, without argument, what I take to be some of the essential features of the shamanic entheogenic cosmology, as it is emerging out of contemporary experiences and reflected in age-old traditions.

- The fundamental reality of the universe is a continuum, a unitive field or fabric, of both energy and consciousness, that is beyond time, space, and all forms, and yet somehow mysteriously within them, simultaneously transcendent and imminent. In traditional Asian religions, this unitive field is variously referred to as *Tao,* or *Atman-Brahman,* or *Tantra* (the "web" or "fabric") or the "jewelled net of Indra." Some Native North Americans refer to it as *Wakan-Tanka,* the all-pervading Creator Spirit. In the traditional Anglo-Saxon religion of the British Isles it was called the *wyrd,* an invisible network of magical forces. In theistic religions

like Christianity, this oneness corresponds to what is called
the Godhead, i.e., beyond the personal deity. In the systems
language of postmodern science it is seen as an infinitely
complex system of interrelationships, or "web of life." At
the level of the planet Earth, this integrated whole is referred
to as Gaia—the name of the ancient Greek Earth Goddess
that has become the name of the whole Earth considered as
a purposive intelligent living superorganism.

- The world or cosmos is multidimensional, a spectrum of
many worlds. In most shamanic traditions we have upper,
middle, and lower worlds. In some mythic-shamanic traditions we have five, seven, nine, or more worlds, often
arrayed around a central tree or axis, the *axis mundi*. Other
names for these nonordinary realms are "spirit world,"
"otherworld," "faery world," and "dreamtime." In esoteric
and thesosophical traditions we usually hear of seven levels
of consciousness, such as the etheric, the astral, the mental,
and so forth. In the Indian and Tibetan traditions as well
there are many levels or realms of consciousness, sometimes
arranged in a circle on a wheel. In the shamanic traditions,
and in the experiences of contemporary neoshamanic practitioners, with or without mind-moving substances, experiences of visiting other worlds are quite common. Also, of
course, they are accessible via dreams. Alternatively, the
person may feel that the veils, barriers, or screens between
worlds can become transparent or porous, so one can see
and be in both the ordinary and the spirit world at the same
time (and in the same place).

Since these other worlds are realms of nonmaterial, transtemporal,
and transspatial consciousness, they are not considered accessible to scientific investigation, and thus not really to exist. The psychiatric term
for perceptions of other realities is "derealization." However, explorers
of consciousness, past and present, report that these other worlds do
indeed exist and they are as real as the familiar material world, in which
we are mostly focused, most of the time. The closest that postmodern

science comes to acknowledging the reality of other worlds is through holistic systems theory, which speaks of multiple levels of wholes and parts. For example, at the universe level we have clusters of galaxies, galaxies, solar systems, and planets. At the planetary level we have biosphere, ecosystems, populations, and species. At the human social level, we have societies, subcultures, organizations, tribes, and families. The organism is composed of organ systems, cells, molecules, atoms and subatomic particles. It is clear that to the degree one expands awareness into these other realms of consciousness, one then lives in a far greater and more encompassing world—one that includes, but is not limited to, the ordinary reality generally supposed to be the only one.

- The universal unitive field or cosmic continuum has a basic symmetrical polarity, referred to by names such as yin and yang, Shiva and Shakti, light and dark, positive and negative charge, male and female, electric and magnetic, Father Sky and Mother Earth and numerous others. These polarities can be observed and experienced at all levels of reality, from the macrocosmic to the microscopic.

- The symmetrically polarized basic continuum differentiates, at all levels, into an infinite variety of names and forms, images and objects, identities and beings. We can recognize this multiplicity at the level of galaxies, stars, and planets; in the biological diversity of plant and animal species on Earth; in the cultural diversity of human societies; and in the psychic multiplicity of our inner life.

- Since we are part of the unified system of interdependence, just like every other being, we can never actually be outside of it, as a detached, objective observer. But since the unified field is energy, we are energetically connected to every other form and being in the universe. And since the field is also consciousness, this enables us, as human beings, to attune with, identify with, and communicate with any and every other life-form, object, or being in the universe, from the macrocosmic to the microscopic.

- Whereas the so-called higher religions associated with literate, urban civilizations tend to be monotheistic, with a single (usually male) deity, the theology of animistic-shamanistic cultures is polytheistic, with an enormous variety in the names and forms of gods and goddesses, particularized for each culture and its mythic tradition. It is not uncommon for participants in sessions with hallucinogenic plants or substances to perceive or feel the presence of deities or spirits from many different cultures, including some with whom they have no genetic, biographical, or geographical connection. Of course, it also happens that people may connect with nature spirits and deities of various traditions without any psychoactive amplification. Recognizing and acknowledging the equal reality of other worlds, shamanic entheogenic explorers also recognize the reality of nonmaterial beings with whom it is possible to communicate—animate essences, living intelligences, "spirits" in the traditional language.

THE ANIMISTIC REVIVAL
AND THE TRANSFORMATION OF SOCIETY

Having presented some of the fundamental features of the animistic, indigenous worldview that is associated with the revival of interest in shamanic practices, including the use of hallucinogens or entheogens, I now want to address the question of what this may mean in the context of the present world situation. What does it mean that people in large numbers are now returning to these ancient traditions of spiritual and healing practice in our world of multinational industrial corporations, of computers and electronic networks?

To return to the argument I proposed in the Introduction, I am saying that the unprecedented industrial-technological assault on the biosphere we are witnessing in our time is rooted in the mechanistic scientism of the modern world, which deliberately divorced itself from spirituality, values, and consciousness. There exists a vast gulf in common understanding between what we regard as sacred and what we regard as natural. And yet, out of the experiences of millions of individuals in the Western world with hallucinogenic sacraments, as well as other shamanic practices, we are seeing the reemergence of the ancient integrative worldview that sees all of life as an interdependent web of relationships that need to be carefully protected and preserved.

The history of the reemergence of hallucinogens and psychoactive plants in the West proceeded in several stages. There are some remarkable synchronicities (C. G. Jung's term for meaningful coincidences) in this history, of which the discovery of LSD is the most dramatic. In 1942, at the height of World War II, the Italian physicist Enrico Fermi, working at the University of Chicago, succeeded in triggering the first nuclear chain reaction, thereby setting the stage for the construction of the first atomic bombs. The power of these bombs exceeded existing explosives by a factor of one thousand. In 1943, the Swiss chemist Albert Hofmann, working with ergot derivatives at Sandoz laboratories in Basel to find treatments for migraine, first accidentally absorbed a tiny amount of lysergic acid diethylamide (LSD). He then tested the drug and found it to be the most potent hallucinogen ever known, exceeding mescaline, the best-known psychoactive at that time, by a factor of one thousand in potency. Thus, in the 1940s, we saw the simultaneous development of atomic energy and a psychoactive drug that acts like an atomic explosion on the human mind, changing forever the worldview and basic life-orientation of all who experienced it.

As the second note in a Gurdjieffian octave of cultural transformations, the decade of the 1950s saw the introduction into the culture of several mind-expanding plant-based shamanic spiritual movements. In 1957, the American banker and mycologist Robert Gordon Wasson rediscovered the sacred mushroom ceremony of the Aztecs under the guidance of the Mazatecan *curandera* Maria Sabina. The publication of his observations in LIFE magazine triggered a surge of experimentation and consciousness exploration in which tens of thousands of young North Americans and Europeans started experimenting with hallucinogenic mushrooms, both in Mexico and elsewhere. Also in the mid-1950s, a Brazilian rubber tapper named Gabriel de Costa, having experienced the hallucinogenic potion ayahuasca, received a vision that he was to start a church in which this "tea" was the central sacrament, the União do Vegetal (UDV), now probably the largest and most tightly organized of the three Brazilian ayahuasca churches. The other two—Santo Daime and Barquinia—also grew and attracted increasing numbers of followers during this period. While separate from the shamanic rituals, the Brazilian ayahuasca churches maintain a respectful and spiritual attitude toward the use of the visionary plant medicines, and a strong feeling of connection to their indigenous roots in shamanic healing practices. The spread of hallucinogenic mushroom use and cultivation connected the psychedelic movement to age-old animistic, shamanistic traditions.

Then, in the 1960s, experiences with consciousness-expanding drugs and plants moved out of the psychiatric clinics and laboratories and triggered a series of profound cultural transformations the full dimensions of which have yet to be fully appreciated. In the early 1960s Timothy Leary and associates began their research with psychedelics at Harvard University, and in 1963 Leary, Metzner, and Alpert published *The Psychedelic Experience—A Manual Based on the Tibetan Book of the Dead*. Around the same time, in California, novelist Ken Kesey and his associates, called The Merry Pranksters, staged a series of rock concerts, called "acid tests," in which thousands of people took LSD while listening to music and watching light shows. Thus was born a revolution in collective consciousness, in which hundreds of thousands of people, perhaps millions, had one or more profound, life-changing psychedelic experiences.

Along with this transformation of collective consciousness, and often involving some of the same people who had experienced psychedelics, the 1960s saw the beginnings or the vitalization of several other sociocultural change movements with profound and lasting impact: the ecology and environmental movement (for which Rachel Carson's 1962 book *Silent Spring* was a major catalyst); an upsurge of creative innovation in music, the arts, fashion, and literature; the women's liberation movement, with its "consciousness-raising" circles (for which Betty Friedan's 1963 book *The Feminine Mystique* was a major catalyst); the sexual revolution and increased freedom of sexual expression, catalyzed by the contraceptive pill; the civil rights, antidiscrimination movement, inspired by Martin Luther King; and the antiwar movement, galvanized by the televised horrors of Vietnam.

In each of these movements, which started in the United States and spread from there throughout the Western-influenced world during the 1970s and 1980s, there was a transcendence, a breaking of what were perceived as the restrictive conventions and social norms of the 1950s and before. This kind of transcendence of conventions, the going beyond the hitherto accepted paradigms of reality and identity that we see in each of these social movements, is basically characteristic of psychedelic and hallucinogenic experiences. It is tempting to speculate whether the introduction of powerful mind-expanding agents, both drug and plant, into the culture, might somehow relate, at some deeper cosmic-karmic level, to the mounting crisis in world civilization.

Certainly, it is not difficult to see the parallels in several cultural movements that seek to correct the dangerous imbalance in humanity's relation to nature: in deep ecology and ecofeminism, which call for a respectful, egalitarian, ecocentric attitude toward the natural world; in the organic gardening and farming movements, which seek to return to traditional methods avoiding chemical fertilizers and pesticides; in the movement to increased use of herbal, nutritional, and complementary healing modalities with less reliance on high-tech interventions; and in several other philosophical, scientific, and religious movements including bioregionalism, ecopsychology, living systems theory, creation spirituality, ecotheology, and others.

In these diverse movements, from many disciplines, to transform our human perceptions, attitudes, and practices in relation to the Earth toward a healthier, nonexploitative, nondominating recognition of interrelatedness, the respectful use of entheogenic plant medicines in spiritual/therapeutic contexts may yet come to play a highly significant role.

In considering the Brazilian ayahuasca churches, some provocative speculations have occurred to several observers. These groups, which along with the Native American Church and the African Bwiti cult, can be considered genuine religious revitalization movements, have expanded their base during the 1980s and 1990s, spreading from Brazil to centers in North America and Europe, and attracting thousands of people. On the face of it, ayahuasca, with its powerful emetic action and sometimes shattering self-revelations, would seem to be an unlikely candidate for a religious sacrament. But it has become just that and has acquired a near-legendary reputation for its healing and empowering attributes. I have myself seen remarkable transformations of personality in people who have become involved in one or another of these churches.

What is happening here? Could these churches become widely popular religions in the twenty-first century? Two thousand years ago, three monotheistic religions arose in the desert borderlands of the Middle East. As the ecologist Paul Shepard has argued, the often harsh and unforgiving environment may have contributed to the idealization of transcendence found in monotheism, as well to its "authoritarian, masculinist and ascetic ideology" that has come to dominate the world stage. In the Brazilian hoasca churches, as well as the Amazonian shamanic traditions from which they originally, though indirectly, derived, the underlying ethos and imagery is very different. Here the essential imagery is of flowing waters and growing plants. The river flows, the inebriating vision-drink flows, the purging vomit flows, the feelings of joy and sadness flow; plant and animal life grows in the luxuriant green abundance of the richest rain forest on Earth. The ultimate theology of these churches is very different. There are hymns, prayers, and invocations of biblical figures, but also the spirits of the forest and the sea, the Sun, Moon, and Stars, and various indigenous deities. This is polytheistic, animistic nature religion, bringing about a reunification of the sacred and the natural.

An even more radical set of questions arises from the visions of some Western and Northern ayahuasqueros, particularly those steeped in evolutionary and ecological biology. Why do so many plants carry psychoactive tryptamines and other chemicals that are capable of producing profound consciousness-transforming perceptions in human beings, opening them up to the deepest mysteries of life and death? On one level this confirms the basic unity of all life on Earth, the oneness of the molecular genetic code. But the usual Darwinian assumption is that nothing evolves by chance—natural selection works to favor those structures and capabilities that are adaptive in some way. So how is it adaptive for plants to produce alkaloids that seemingly serve no other particular function, and yet provide profound healing or insight in the human?

There seems to be some kind of strange symbiosis going on. We know there are many aspects of what has been called "the great symbiosis" between the plants and the animals in the biosphere of Earth. First, there is the constant invisible worldwide exchange of gases: the oxygen breathed out by the plants is nourishment for the animals, and the CO_2 emitted by the animals is breathed in and converted by the green plants. At the level of fruit-bearing bushes and trees, the symbiosis is more visible: plants produce fruit, which are seed packages. Animals eat the fruit, and the seed kernels inside get propagated some distance away, where they will have more space to grow. So we animals are basically working for the plants, one could say, as seed carriers. But it is a fair and even abundant exchange: most of our nourishment, most of our medicine and healing, all our tonics and extracts for well-being and longevity come from the plant realm.

So with these "plant teachers," as ayahuasqueros call them, there must also be an exchange. We get knowledge, insight, psychic or physical healing from the plant teachers. In exchange, we should give something back. Individuals who have found themselves at this juncture may at first not know how or what to give back. If they then ask the plant teachers, or ask themselves, how do we give back, how do we repay what appears to be a gift of astounding generosity from the plant teachers, the answers are remarkably consistent. They have to do, as one might expect, with practices that reduce our adverse impact on the ecosystems, and with the preservation of wilderness and the essential diversity of life. That's why so many people who have experienced ayahuasca (as well as other psychedelics, other shamanic practices, near-death experiences, or the death of a loved one) become deeply involved in ecological preservation and sustainability projects, as well as in efforts to preserve the culture of indigenous peoples.

There may be a profound and mysterious shift occurring in the balance of life on this planet. The dominant and dominating role of the human in relation to the natural world has brought about unparalleled ecological disaster, degradation of habitats, and loss of species. Could it be that the profound consciousness-raising and compassion-deepening effects of the visionary plant brews and tinctures are signaling an evolutionary initiative coming from other, nonhuman, intelligences on this planet? Instead of the usual attitude of arrogant and exploitative superiority, those who have experienced ayahuasca and other entheogens are more likely to find themselves humbled and awed by the mysterious powers of nature, and strive to live in a simpler way that minimizes environmental harm and celebrates the astonishing diversity and beauty of life.

ᑎotes ᴑᴎ Cᴑᴎtribᴜtᴑrs

Ralph Metzner, Ph.D., obtained a B.A. in philosophy and psychology at Oxford University, and a Ph.D. in clinical psychology at Harvard University; he also held a postdoctoral fellowship in psychopharmacology at the Harvard Medical School. He worked with Timothy Leary and Richard Alpert on psychedelic research, edited the *Psychedelic Review*, coauthored *The Psychedelic Experience* (1964), and edited *The Ecstatic Adventure* (1968). He is also the author of *Maps of Consciousness* (1971), *Know Your Type* (1979), *Opening to Inner Light* (1986), *The Well of Remembrance* (1994), *The Unfolding Self* (1998), *Green Psychology* (1999), and *Sacred Mushrooms of Visions: Teonanácatl* (2005). He has pursued research in altered states of consciousness and cross-cultural methods of consciousness expansion and published more than one hundred articles on consciousness, shamanism, alchemy, transformation, and mythology. He is a professor of psychology at the California Institute of Integral Studies in San Francisco and maintains a private practice of psychotherapy in the Bay Area. He is president and cofounder of the Green Earth Foundation, a nonprofit educational organization devoted to healing and harmonizing the human relationship with the Earth. Dr. Metzner can be contacted via e-mail at: ralph@greenearthfound.org. His Web site address is: www.greenearthfound.org.

Charles S. Grob, M.D., is Director of the Division of Child and Adolescent Psychiatry at Harbor-UCLA Medical Center and Professor of Psychiatry and Pediatrics at the UCLA School of Medicine. He did his undergraduate work at Oberlin College and Columbia University and obtained a B.S. from Columbia in 1975. He obtained his M.D. from the State University of New York, Downstate Medical Center in 1979. Prior to his appointment at UCLA, Dr. Grob has held teaching and clinical positions at the University of California, Irvine, College of Medicine and The Johns Hopkins University School of Medicine, Departments of Psychiatry and Pediatrics. He has conducted thus far the only government-approved psychobiological studies of MDMA, as well as participating, with McKenna, Callaway, and others, in the Brazilian hoasca project. He has published numerous articles on his research in medical and psychiatric journals and collected volumes. He is a founding board member of the Heffter Research Institute. Dr. Grob can be reached by e-mail at: grob@humc.edu.

Dennis J. McKenna, Ph.D., received a Master's Degree in Botany from the University of Hawaii in 1979, and his Ph.D. in Botanical Sciences from the University of British Columbia in 1984. His doctoral research focused on a phytochemical and pharmacological investigation of Amazonian psychoactive plants. Following the completion of his doctorate, Dr. McKenna received postdoctoral research fellowships in the Laboratory of Clinical Pharmacology, National Institute of Mental Health, and in the Department of Neurology, Stanford University School of Medicine. He is coauthor, with his brother Terence, of *The Invisible Landscape: Mind, Hallucinogens, and the I Ching* (1975; Citadel Press, 1991), a speculative metaphysical exploration of the ontological implications of psychedelic drugs that resulted from the two brothers' early investigations of Amazonian hallucinogens. He worked for Shaman Pharmaceuticals as Director of Ethnopharmacology and for the Aveda Corporation, in Minnesota, a manufacturer of natural cosmetic products, as Senior Research Pharmacognosist. He currently works as a scientific consultant to clients in the herbal, nutritional, and pharmaceutical industries. Dr. McKenna is author or coauthor on over thirty scientific papers in journals such as *Journal* *of Ethnopharmacology, European Journal of Ethnopharmacology,* and others. He is a founding board member and Vice-President of the Heffter Research Institute and also serves as board member and Research Advisor to Botanical Dimensions Foundation. Dr. McKenna can be reached at: djmckenna@aol.com.

J. C. Callaway, Ph.D., obtained his B.Sc. in chemistry in 1980 from Henderson University in Arkansas, a Master's Degree (1984) in organic chemistry from the University of Mississippi, in Oxford, Mississippi, and a Ph.D. (1994) in medical chemistry from the University of Kuopio, in Kuopio, Finland. His primary research focus has been medicinal chemistry, pharmacognosy, and psychopharmacology, and especially in the quantitative analysis of alkaloids in plants, humans, and other animals. He has published extensively on the chemistry and pharmacology of ayahuasca and other tryptamines, as well as the chemistry of hemp, in scientific, peer-reviewed journals such as the *Journal of Ethnopharmacology, Pharmacology and Toxicology, Journal of Analytical Toxicology, Journal of Nervous and Mental Disease, Journal of Psychoactive Drugs, Planta Medica, Human Psychopharmacology,* and *American Journal of Psychiatry.* He is the author of the monograph *Pinoline and Other Tryptamine Derivatives* (1994). He currently holds a research appointment in the Department of Pharmaceutical Chemistry and teaches in the Department of Pharmacology and Toxicology, both at the University of Kuopio, Finland. He can be reached via e-mail at: callaway@uku.fi.